Rapid Assessment Program

A Rapid Biological Assessment of the Upper Palumeu River Watershed (Grensgebergte and Kasikasima) of Southeastern Suriname

Editors:
Leeanne E. Alonso and Trond H. Larsen

RAP

Bulletin
of Biological
Assessment

67

CONSERVATION INTERNATIONAL - SURINAME

CONSERVATION INTERNATIONAL

GLOBAL WILDLIFE CONSERVATION

ANTON DE KOM UNIVERSITY OF SURINAME

THE SURINAME FOREST SERVICE (LBB)

NATURE CONSERVATION DIVISION (NB)

FOUNDATION FOR FOREST MANAGEMENT AND PRODUCTION CONTROL (SBB)

SURINAME CONSERVATION FOUNDATION

THE HARBERS FAMILY FOUNDATION

The RAP Bulletin of Biological Assessment is published by:

Conservation International
2011 Crystal Drive, Suite 500
Arlington, VA USA 22202

Tel : +1 703-341-2400
www.conservation.org

Cover photos:
The RAP team surveyed the Grensgebergte Mountains and Upper Palumeu Watershed, as well as the Middle Palumeu River and Kasikasima Mountains visible here. Freshwater resources originating here are vital for all of Suriname. (T. Larsen)

Glass frogs (*Hyalinobatrachium* cf. *taylori*) lay their eggs on leaves overhanging flowing streams, and are an indicator of the high water quality of the Upper Palumeu River Watershed. (S. Nielsen)

The Upper Palumeu River Watershed supports a high diversity of fishes, including this cichlid (*Guianacara owroewefi*). (T. Larsen)

Editors: Leeanne E. Alonso and Trond H. Larsen

Maps: Jesus Aguirre, Olaf Bánki, Luis Barbosa, Krisna Gajapersad, Priscilla Miranda, Sara Ramirez, and Leonardo Sáenz

ISBN: 978-1-934151-57-0

Suggested citation:
Alonso, L.E and T.H. Larsen (eds.). 2013. A Rapid Biological Assessment of the Upper Palumeu River Watershed (Grensgebergte and Kasikasima) of Southeastern Suriname. RAP Bulletin of Biological Assessment 67. Conservation International, Arlington, VA.

This RAP survey and publication of this RAP report were made possible by generous financial support from The Suriname Conservation Foundation.

Table of Contents

RAP SCIENTIFIC TEAM

Leeanne E. Alonso, PhD (RAP Team Leader, Ants)
Global Wildlife Conservation
P.O. Box 129
Austin, TX, 78767-0129 USA

Olaf Bánki, PhD (Plants)
Netherlands Centre for Biodiversity Naturalis
National Herbarium of the Netherlands
P.O. Box 9514, 2300 RA Leiden,
The Netherlands

Chequita Bhihki, MSc (Plants)
Netherlands Centre for Biodiversity Naturalis
National Herbarium of the Netherlands
P.O. Box 9514, 2300 RA Leiden,
The Netherlands

Krisna Gajapersad, BSc (Large Mammals)
Conservation International – Suriname
Kromme Elleboogstraat 20
Paramaribo, Suriname

Jackson Helms, MSc (Ants)
Department of Biology
University of Oklahoma
730 Van Vleet Oval, Room 314
Norman, OK 73019 USA

Rawien Jairam (Reptiles and Amphibians)
National Zoological Collection of Suriname
Anton de Kom University of Suriname
Leysweg 9
Paramaribo, Suriname

Gwendolyn Landburg, MSc (Water Quality)
National Zoological Collection of Suriname/Center for
Environmental Research
Anton de Kom University of Suriname
Leysweg 9
Paramaribo, Suriname

Trond H. Larsen, PhD (Dung Beetles)
Rapid Assessment Program
Conservation International
2011 Crystal Drive, Suite 500
Arlington, VA 22202 USA

Burton K. Lim, PhD (Small Mammals)
Department of Natural History
Royal Ontario Museum
100 Queen's Park
Toronto, ON M5S 2C6 Canada

Jan Mol, PhD (Fishes)
CELOS
Prof. Dr. Ruinardlaan
Paramaribo, Suriname

Piotr Naskrecki, PhD (Katydids)
Museum of Comparative Zoology
Harvard University
26 Oxford Street
Cambridge MA 02138 USA

Stuart Nielsen, MSc (Reptiles and Amphibians)
Department of Biology
University of Mississippi
Box 1848
University, MS 38677 USA

Brian J. O'Shea, PhD (Birds)
North Carolina Museum of Natural Sciences
11 W. Jones St.
Raleigh, NC 27601 USA

Serano Ramcharan (Birds)
Cayottestraat 43
Paramaribo, Suriname

Andrew Short, PhD (Aquatic Beetles)
Department of Ecology & Evolutionary Biology and Biodiversity Institute
University of Kansas
1501 Crestline Drive, Suite 140
Lawrence, KS 66045 USA

Arioene Vreedzaam, MSc (Primates)
Proffessor Dr. Ruinardlaan 5
ADEK Campus
Paramaribo, Suriname

Kenneth Wan Tong You (Fishes)
CELOS
Prof. Dr. Ruinardlaan
Paramaribo, Suriname

ADDITIONAL RAP TEAM MEMBERS

Priscilla Miranda (Coordination)
Conservation International – Suriname
Kromme Elleboogstraat 20
Paramaribo, Suriname

Ted Jantz (Videographer and expert Fisherman)
Media Vision
Paramaribo, Suriname

Rafael Jantz (Videographer)
Media Vision
Paramaribo, Suriname

Andre Semmie (Tree Spotter)
Conservation International – Suriname
Kromme Elleboogstraat 20
Paramaribo, Suriname

Jeffrey Krimbo (Camera Trapping and Plants)
Conservation International – Suriname
Kromme Elleboogstraat 20
Paramaribo, Suriname

Hermando Banda (Small Mammals)
Conservation International – Suriname
Kromme Elleboogstraat 20
Paramaribo, Suriname

Priscilla Dragtenstein (Game Warden)
LBB/NB Nature Conservation Division
Cornelis Jongbawstraat 12
Paramaribo Suriname

Fabian Lingaard (Game Warden)
LBB/NB Nature Conservation Division
Cornelis Jongbawstraat 12
Paramaribo Suriname

Richard Conniff (Journalist)
Connecticut, USA

Randy Olson (Photographer)
Pennsylvania, USA

Russell A. Mittermeier, PhD (President)
Conservation International
2011 Crystal Drive, Suite 500
Arlington, VA 22202 USA

Richard Sneider, PhD (RAP Supporter)
Conservation International: Chairman's Council
Greater Los Angeles Zoo: Board member and Trustee
HDI: Board Member
Los Angeles, CA, USA

Fabian Oberfeld (RAP Supporter)
Conservation International: Chairman's Council
Los Angeles, CA USA

Jeffrey Kapor (RAP Supporter)
Los Angeles, CA USA

FIELD GUIDES

From Apetina	From Palumeu
Malasijana Ukili	Vincent Wedije
Johan Neni	Remigio Mitian
Hendrik Neni	Atei Arekawa
Lesley Kawaidoe	Kelomeu Nailoepum
Sipali Mettelli	Michael Ankarapi
Pensaman Ukili	Jaka Tawaike
Saloman Koemapu	Sosi Makainoe
Gilbert Koemaja	Tabo Tetijesë
Rene Koemapu	James Akinare
Eipu Ajamaka	Gilbert Alakoeng
Jula Meliwa	Patrik Ikoewa
Jakopi Ikoewa	Masoe Tetijesë
Alamoike Ukili	Pajanoe
Mattiju Koemapu	Keesi
Richard Kawaidoe	
Lucien Meliwa	

CONSERVATION INTERNATIONAL SURINAME

Conservation International Suriname is an environmental organization that has worked in country for the last 20 years to protect nature for the benefit of people. CI-Suriname is focused on supporting national policy for green economic development by supporting land use planning and demonstrating the value of ecosystem services in Suriname.

Conservation International Suriname
Kromme Elleboogstraat no. 20
Paramaribo
Suriname
Tel: 597-421305
Fax: 597-421172
Web: http://www.conservation.org/where/south_america/
suriname/Pages/suriname.aspx

CONSERVATION INTERNATIONAL

Conservation International (CI) is an international, non-profit organization based in Arlington, VA. CI believes that the Earth's natural heritage must be maintained if future generations are to thrive spiritually, culturally and economically. Building upon a strong foundation of science, partnership and field demonstration, CI empowers societies to responsibly and sustainably care for nature, our global biodiversity, for the well-being of humanity.

Conservation International
2011 Crystal Drive, Suite 500
Arlington, VA 22202
Tel: (703) 341-2400
Web: www.conservation.org

GLOBAL WILDLIFE CONSERVATION

Global Wildlife Conservation (GWC) protects endangered species and habitats through science-based field action. GWC conserves the world's most endangered species and their habitats through exploration, research, and conservation. Our dynamic model combines a central science-based strategy with local action in partnership with our in-country collaborators. We are a revolutionary organization that builds upon the collective knowledge of conservation pioneers with modern methods and tools, centered on the urgency for action to prevent species extinctions.

Global Wildlife Conservation
P.O. Box 129
Austin, TX 78767-0129, USA
Web: www.globalwildlife.org

ANTON DE KOM UNIVERSITY OF SURINAME

Anton de Kom University of Suriname was founded on 1 November 1968 and offers studies in the fields of social, technological and medical sciences. There are five research centers conducting research and rendering services to the community. The Center for Agricultural Research (CELOS) is promoting agricultural scientific education at the faculty of Technological Sciences, Institute for Applied Technology (INTEC), Biomedical Research Institute, Institute for Development Planning and Management (IDPM), Institute for Research in Social Sciences (IMWO), the Library of ADEK, University Computer Center (UCC), National Zoological Collection (NZCS) and National Herbarium of Suriname (BBS).

The primary goals of the NZCS and BBS are to develop an overview of the flora and fauna of Suriname, and build a reference collection for scientific and educational purposes. The NZCS also conducts research on the biology, ecology, and distribution of certain animal species or on the composition and status of certain ecosystems.

Anton de Kom University of Suriname
Universiteitscomplex / Leysweg 86
P.O. Box 9212
Paramaribo
Suriname
Web: www.uvs.edu

National Zoological Collection of Suriname (NZCS)
Universiteitscomplex / Leysweg 9
Building # 17
P.O. Box 9212
Paramaribo
Suriname
Tel: 597-494756
Email: nzcs@uvs.edu

National Herbarium of Suriname (BBS)
Universiteitscomplex / Leysweg
Paramaribo
Suriname
Tel: 597-465558
Email: bbs@uvs.edu

THE SURINAME FOREST SERVICE (LBB/NB/SBB)

The Suriname Forest Service (LBB) within the Ministry of Physical Planning, Land and Forest Management is responsible for forest management and nature conservation. One of its agencies is the Nature Conservation Division (NB), which is assigned the task of managing all protected areas. As the CITES authority in Suriname, it also issues all permits for research and for the export of species, as well as enforcement within the framework of the laws on hunting and wildlife. The second agency is the Foundation for Forest Management and Production Control (SBB), which is mandated to manage production forests and is responsible for supervision and control of all logging.

The Suriname Forest Service (LBB) and
Nature Conservation Division (NB)
Cornelis Jongbawstraat # 10 - 12
Paramaribo, Suriname

Foundation for Forest Management and Production Control (SBB)
Ds. Martin Luther Kingweg perc. no. 283
Paramaribo, Suriname

SURINAME CONSERVATION FOUNDATION

Created in March 2000, by effective cooperation between the Government of Suriname and Conservation International, the Suriname Conservation Foundation (SCF) is the Surinamese Environment Fund par excellence, committed to the protection of Suriname's biodiversity, and protected areas in particular. Through its strong institutional framework, sound financial mechanism, good national partnerships and international recognition, strong cooperation with stakeholders, and a visible commitment to the sustainable utilization and protection of nature, SCF makes recordable contributions; both by project financing and strategic actions that promote national economic development and the protection and preservation of the biodiversity of our planet.

Suriname Conservation Foundation (SCF)
Dr. J.F Nassylaan # 17
Paramaribo-Suriname (South America)
Tel: 597-470155
Fax: 597-470156
E-mail: surcons@scf.sr.org
Web: www.scf.sr.org

THE HARBERS FAMILY FOUNDATION

Jeff Harbers loved the outdoors and felt called to protect and preserve the earth's natural beauty and diversity for future generations. As an early Microsoft employee, he had the resources to think big. In 1998, Jeff made a strategic contribution to Conservation International and helped create the Central Suriname Nature Reserve (CNSR), one of the largest nature reserves in South America, and arguably the most pristine protected tropical forest on the planet. In 2006, Jeff was killed in a tragic plane crash, but his legacy lives on through the work of the Harbers Family Foundation (HFF). The success of the CSNR prompted the HFF to increase its level of commitment and participate as a partner with CI to develop the South Suriname Conservation Project, currently underway.

The Harbers Family Foundation brings human and environmental issues into focus through the creation of powerful visual narratives designed to inspire change in attitudes, beliefs and behaviors. Working with some of the world's leading photographers and visual storytellers, the Foundation highlights the diverse initiatives of cutting edge nonprofits and NGOs around the world, with an eye on global conservation issues and humanitarian need.

The Harbers Family Foundation
Web: www.harbersfoundation.org

Acknowledgments

The RAP team extends our deepest gratitude to Captain Peciphe Padoe and the people of Palumeu for granting us permission to study the biodiversity of their lands. We especially wish to thank the 16 men from Apetina and the 14 men from Palumeu who accompanied the RAP team in the field and provided invaluable service piloting the boats around and through dangerous rapids, setting up two first-rate field camps, cooking meals, assisting and guiding the team in the forest, and for overall outstanding support to the expedition. Special thanks to Johan Neni for his leadership in the field and to Ted Jantz for helping to link the RAP expedition with the local communities.

We are especially grateful to the Suriname Conservation Foundation for their generous financial support of this RAP survey and for their continued support of conservation and scientific capacity building in Suriname. We thank One World Apparel LLC, Weavers, and Band of Gypsies for their generous financial support of the RAP survey.

We also thank Global Wildlife Conservation, The Harbers Family Foundation, the National Zoological Collection of Suriname of the Anton de Kom University of Suriname, and CELOS for partnering with CI-Suriname on this expedition.

We thank Natascha Veninga and Letitia Kensen of CI-Suriname who provided logistical assistance at all stages of the RAP, especially for the helicopters and flights. Armand Moredjo provided overall technical guidance for the RAP and helped to obtain necessary permits. Annette Tjon Sie Fat, Theo Sno, and Roald Tjon also provided essential support and guidance. We especially thank Krisna Gajapersad who expertly guided all logistics in the field, including the advance recon team and set up of the two field camps, and got the RAP team smoothly out of the field around and through many rapids.

Gregg Gummels and the staff of Gum Air ensured safe transportation of our personnel, equipment, and supplies to and from Palumeu. HiJet provided excellent helicopter service, which got us to RAP Camp 1 and into the Grensgebergte Mountains. We especially thank Edward Geerlings, Reqillio Courtar, and Julio Ladino Santillon for their patience, flexibility and friendly flying.

We thank Armand Bhagwandas and METS for accommodating us in Palumeu, for helping to store and manage our equipment, and for being willing to accommodate the diverse needs of our large team.

We are grateful to the personnel of LBB/NB (Nature Conservation Division), especially Claudine Sakimin, for support in granting us the proper research and export permits. We thank the National Zoological Collection of Suriname and the National Herbarium of Suriname (Anton de Kom University of Suriname) for assistance with the permit application process.

We thank Medische Zending for providing two medics to attend to any medical needs of the RAP team. Special thanks to Broeder Erwin Makosi and Broeder Evany Adipi for their help in the field.

The RAP botanical team gratefully acknowledges the support of the Suriname Forest Control Foundation (SBB), CELOS, and the National Herbarium of Suriname (BBS). Special thanks to Priscilla Dragtenstein and Fabian Lingaard from the LBB/NB Nature Conservation Division for assisting the RAP team. We thank the Alberta Mennega Stichting for funding the travel costs of C. Bhikhi for the determination of the plant specimens in the National Herbarium of the Netherlands at the University of Leiden.

The Ant team thanks Jeffrey Sosa-Calvo (Smithsonian) for assistance in species identifications and Olaf Bánki for identification of the ant-plants.

The Water Beetle team thanks Vanessa Kadosoe (University of Suriname) for assistance in sorting and identifying material. Kelly Miller (University of New Mexico) and Crystal Maier (University of Kansas) provided helpful assistance with identifications of the Dytiscidae and Dryopoidea respectively. Sarah Schmits provided key databasing assistance and support. Trond Larsen thanks Fernando Vaz-de-Mello for assistance with some dung beetle identifications.

The Herpetology team would specifically like to thank Antoine Fouquet and Philippe Kok for assistance in identifying some of the more taxonomically confusing species. We also wish to thank the other members of the RAP team—both scientists and guides/members of the support team—that contributed to this study via opportunistic collections and/or photographic vouchers.

The Bird Team would like to acknowledge Greg Budney and Matt Medler of the Macaulay Library, Cornell Lab of Ornithology, for the generous loan of recording equipment used during this survey.

The Fish Team thanks the people of Palumeu and Apetina villages for acting as guides, safely piloting the boats up and down rapids, talking about their fishing experiences, and in addition catching some fishes with hook-and-line. Other RAP participants also helped us catch specimens, including R. Jairam, T. Jantz, R. Jantz and T. Larsen. We thank the Smithsonian Institution and Dr. Richard Vari for financing a visit to the National Museum of Natural History (Smithsonian Institution), Washington D.C., which allowed us to improve on the quality of identification of the collected fish taxa. Fernando Jerep helped with identification of and other small characoids, Richard Vari checked the identity of, and Philip Willink helped with the identification of some small-sized catfish species. Pierre Yves Le Bail informed us about the occurrence of and aff. in French Guiana. Sonia Fisch-Muller identified an early stage of the ancistrine catfish.

Report at a Glance

DATES OF RAP SURVEY

March 8–29, 2012

DESCRIPTION OF RAP SURVEY SITES

The RAP team surveyed the aquatic and terrestrial flora and fauna of the Upper Palumeu River Watershed of Southeastern Suriname, close to the border with Brazil. This area is possibly the most remote, pristine, and unexplored rainforest region of Suriname. The area is entirely forested with 16 different land cover types ranging from lowland floodplain forest to isolated mountain peaks over 780 m elevation. The RAP team surveyed four sites: 1) Upper Palumeu River, where the river becomes a small creek, mostly seasonally flooded forest with some high dryland forest on granite hills and swamp forest at ~270 m a.s.l., 2) Grensgebergte mountaintop at 800 m, exposed granitic rock surrounded by forest, with a mixture of vegetation types including cyper grasses and bromeliads with orchids and gesneriads on the slopes, low shrub vegetation on the rock, low savannah forest and dryland forest on granite hills, 3) Makrutu Creek at the junction of the Upper Palumeu River and the Makrutu Creek, aquatic ecosystems and seasonally flooded forest along the waterways, and 4) Middle Palumeu River and Kasikasima Mountain, a unique granitic mountain formation that rises over 700 m above the rainforest with vegetation similar to that of the Grensgebergte, and lowland (~200 m a.s.l.) seasonally flooded forest, high dryland forest on granite hills, and savanna forest.

REASONS FOR THE RAP SURVEY

Southeastern Suriname is one of the last extensive, pristine tracts of rainforest left on Earth. Conservation of these forests and rivers, and the natural services they provide to the people of Suriname is important to the future of the country and the region. Since virtually nothing is known scientifically, the first step in protecting Southeastern Suriname is to collect baseline biological and socio-economic data for the region. This RAP survey provides data to guide conservation and sustainable development activities in Southeastern Suriname and provide the scientific justification for protection of this diverse and important region.

MAJOR RESULTS

Results from all of the taxonomic groups surveyed during the RAP survey reveal that Southeastern Suriname contains very high biodiversity and is in pristine condition with virtually no human influence. All of the taxonomic groups except the large mammals indicate that Southeastern Suriname is unique from other areas of the Guiana Shield, containing many species not found elsewhere. Plant species composition differs from the flora of northern Suriname and several bird species appear in Southeastern Suriname that are not found in the north. Water quality and fish diversity are high, indicating that the area, which encompasses the headwaters of many of Suriname's major rivers, provides plentiful freshwater resources. The range of elevations within the mountain ranges and the pristine nature of the lowland forests within Southeastern Suriname contribute to the high biological diversity of the region. We found over fifty species that are probably new to science, including eleven fishes, six frogs, one snake, and many insects. The RAP results highlight the importance of the diversity of the forests, species, and watersheds of Southeastern Suriname.

Species recorded during the Southeastern Suriname RAP survey, March 2012

Taxon	Camp 1 Upper Palumeu (Juuru) 277 m	Camp 2 Grensgebergte Mountain 790–820 m	Camp 3 Makrutu River Camp 240–260 m	Camp 4 Kasikasima 201 m	Total (all 3 camps)	# species potentially new to science	# species new records for Suriname
Plants	161	68	27	161	354		15
Aquatic Beetles	92	10	-	105	157	26	TBD*
Dung Beetles	93	40	-	74	107	10	TBD
Katydids	29	2	-	34	52	6	
Ants	72	25	-	92	149	TBD	TBD
Fishes	71	-	16	49	94	11	2–4
Reptiles	26	7	-	21	42	1	
Amphibians	30	6	-	24	47	6	
Birds	196	103	-	233	313		
Small Mammals	25	12	-	23	39		
Large Mammals	16	-	-	20	24		
Total	811	273	43	836	1378	60	17–19

*TBD = To be determined.

Species of Conservation Concern documented during the 2012 RAP survey of Southeastern Suriname

IUCN Red List of Threatened Species categories: VU=Vulnerable, CR=Critically Endangered (see www.redlist.org); CITES (Convention on International Trade of Endangered Species and Wild Flora and Fauna) Appendices I, II, and III (see www.cites.org)

Group	Species	IUCN or CITES category
Plants		
	Vouacapoua americana	CR
	Syagrus stratincola	VU
Orchidaceae	Cleistes rosea	CITES Appendix II
Orchidaceae	Dichaea picta	CITES Appendix II
Orchidaceae	Epidendrum densiflorum	CITES Appendix II
Orchidaceae	Epidendrum nocturnum	CITES Appendix II
Orchidaceae	Maxillaria discolor	CITES Appendix II
Orchidaceae	Phragmipedium lindleyanum	CITES Appendix II
Birds		
	Pipile cumanensis	VU
	Crax alector	VU
	Myrmotherula surinamensis	VU
	Patagioenas subvinacea	VU
Mammals		
	Pteronura brasiliensis	EN
	Tapirus terrestris	VU
	Ateles paniscus	VU

CONSERVATION RECOMMENDATIONS

Southeastern Suriname is a global natural treasure. There are very few places left on Earth that are as pristine and untouched as this region. The region contains high diversity of plants and animals and a unique composition of species that sets it apart from other areas of Suriname, the Guiana Shield, and the world. Southeastern Suriname contains the headwaters of some of the largest rivers of Suriname that provide clean drinking water and food sources for the people of Suriname, as well as supporting downstream agriculture and energy production. The region's intact forests provide a continuing source of food, medicines and building materials for local people, as well as regulating regional and global climate. Our findings also show that Southeastern Suriname will be disproportionately important for providing freshwater resources under future climate change scenarios.

We strongly recommend protection of Southeastern Suriname to preserve its unique and diverse species and freshwater resources for the people of Suriname and the world, today and for generations to come.

Maps and Photos

Map 1. Suriname and its protected areas. This RAP survey focused on the Grensgebergte Mountains in the South, as well as Kasikasima Mountain. Other mountain ranges of Southeastern Suriname remain almost completely unknown to scientists. Map by Sara Ramirez.

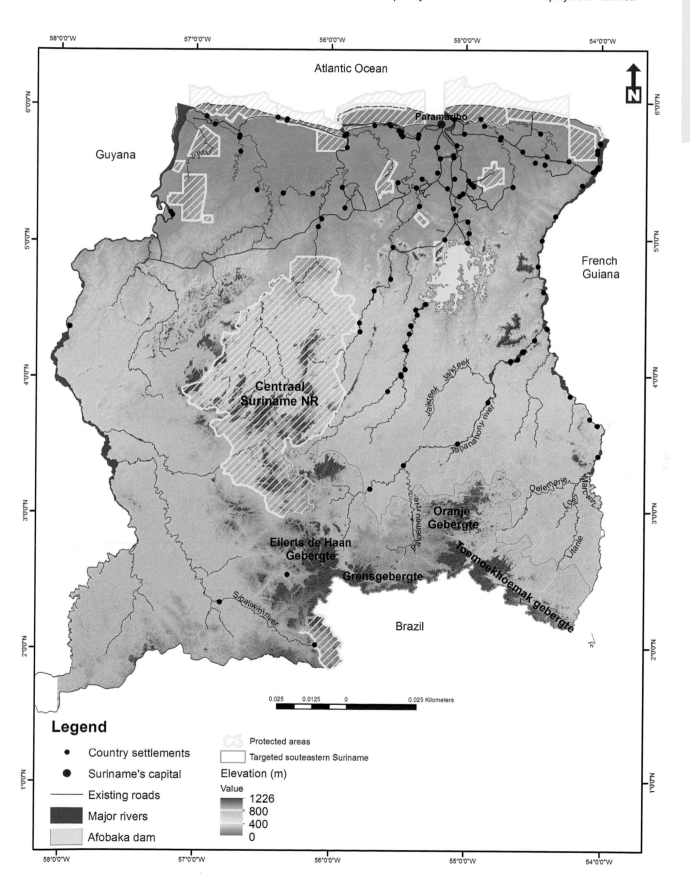

Map 2. Map of RAP survey route and camp sites. Color indicates topography (orange/red high); diagonal lines across rivers indicate rapids (many of which required porting boats through the forest). Map by Priscilla Miranda.

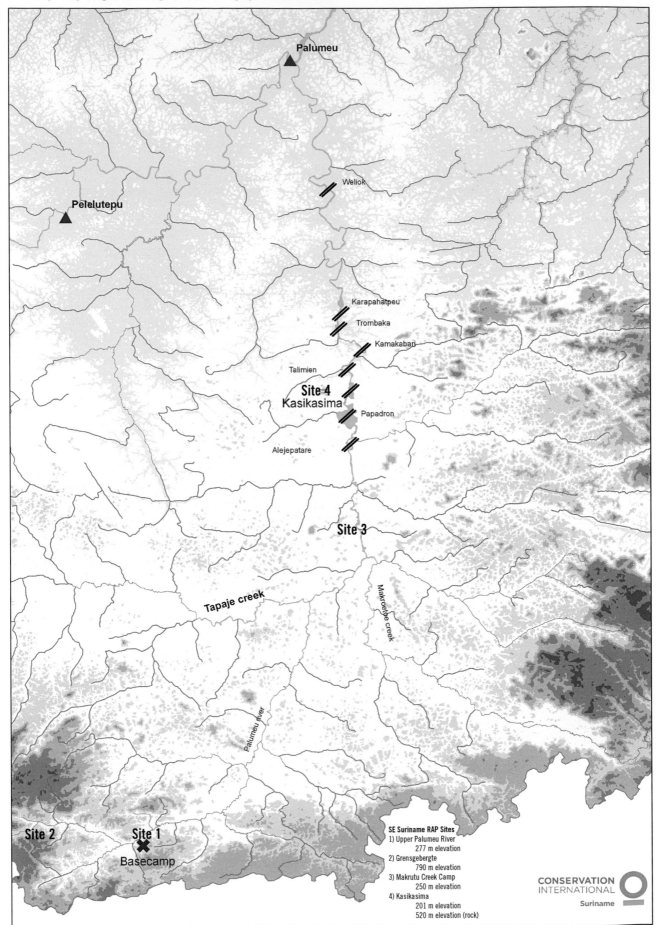

Palumeu

Pelelutepu

Weliok

Karapahatpeu

Trombaka

Kamakabari

Talimien

Site 4
Kasikasima

Papadron

Alejepatare

Site 3

Makroebe creek

Tapaje creek

Palumeu river

Site 2

Site 1
Basecamp

SE Suriname RAP Sites
1) Upper Palumeu River
   277 m elevation
2) Grensgebergte
   790 m elevation
3) Makrutu Creek Camp
   250 m elevation
4) Kasikasima
   201 m elevation
   520 m elevation (rock)

CONSERVATION
INTERNATIONAL
Suriname

Map 3. Vegetation map of Southern Suriname. Map by Jesus Aguire and Olaf S. Banki.

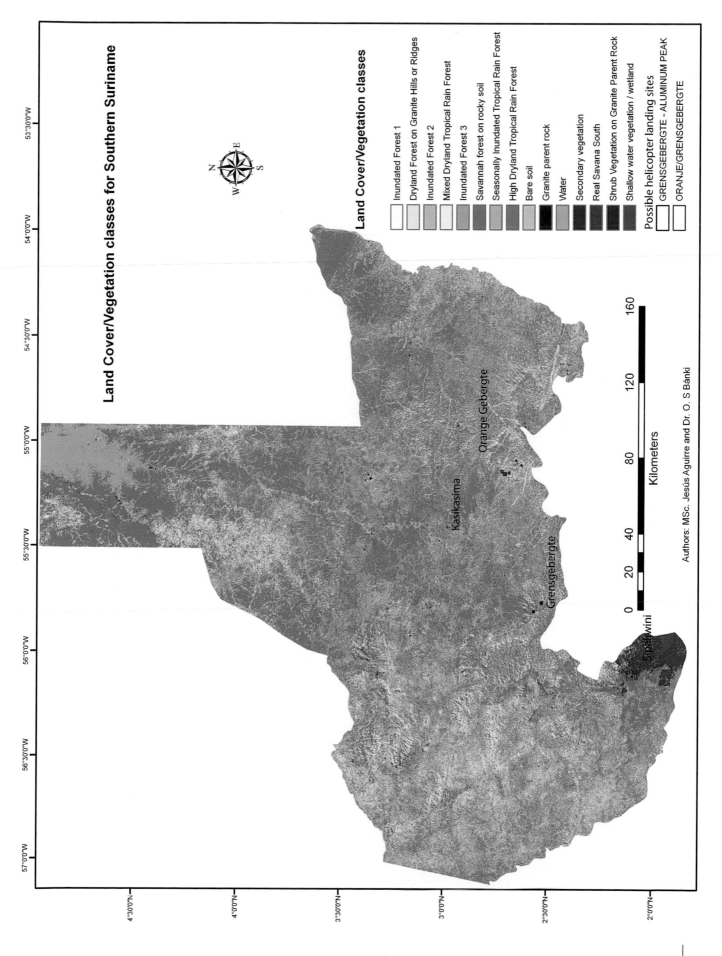

Land Cover/Vegetation classes for Southern Suriname

Land Cover/Vegetation classes

- Inundated Forest 1
- Dryland Forest on Granite Hills or Ridges
- Inundated Forest 2
- Mixed Dryland Tropical Rain Forest
- Inundated Forest 3
- Savannah forest on rocky soil
- Seasonally Inundated Tropical Rain Forest
- High Dryland Tropical Rain Forest
- Bare soil
- Granite parent rock
- Water
- Secondary vegetation
- Real Savana South
- Shrub Vegetation on Granite Parent Rock
- Shallow water vegetation / wetland

Possible helicopter landing sites
- GRENSGEBERGTE - ALUMINUM PEAK
- ORANJE/GRENSGEBERGTE

Orange Gebergte

Kasikasima

Grensgebergte

Sipaliwini

Kilometers

0 20 40 80 120 160

Authors: MSc. Jesús Aguirre and Dr. O. S Bánki

Map 4. Natural resource use map demonstrating the importance of ecosystem services for communities in Southeastern Suriname. Colors ranging from green to red illustrate the relative overall value of ecosystem services. Landscape features, such as rivers and mountains, and a diversity of habitat types provide a range of important services people depend upon (orange and red areas). Map by Sara Ramirez.

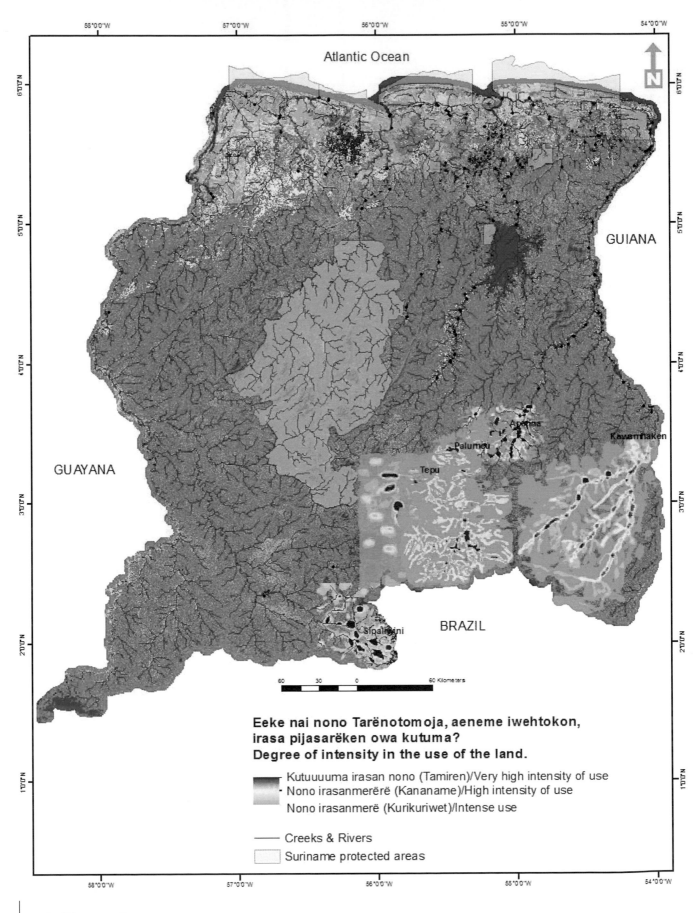

Map 5. Predicted impacts of climate change on water resources in Suriname. 1 km change in water yield for Suriname (mm year-1) using an average of all climate models for the a2 emissions scenario and using the model FIESTA /WaterWorld (Mulligan 2012; Saenz 2012a). Climate resilience of water resources ranges from low (yellow), especially in western Suriname, to high (blue) in eastern Suriname. Southeastern Suriname, with its multiple mountain ranges, is one of the most climate resilient parts of Suriname and will be disproportionately important for supplying water to the country in the future (Sáenz 2012b) (for list of references, see Chapter 1). Map by L. Saenz.

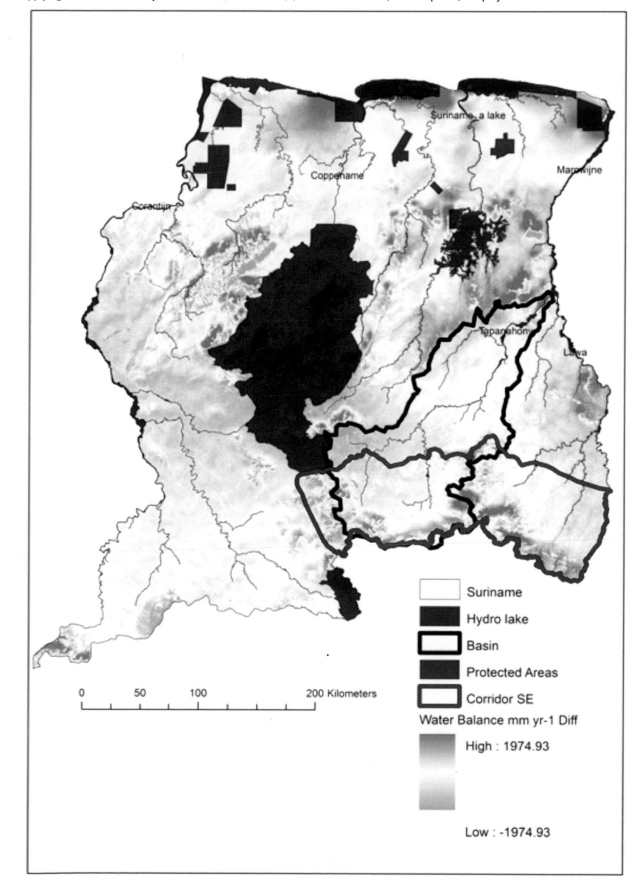

Map 6. Protected areas and indigenous lands of the Guiana Shield. Southeastern Suriname is highlighted in orange, and represents a key wilderness area for connecting and amplifying the effectiveness of neighboring protected areas and indigenous lands. Map by Luis Barbosa (CI-Brazil)

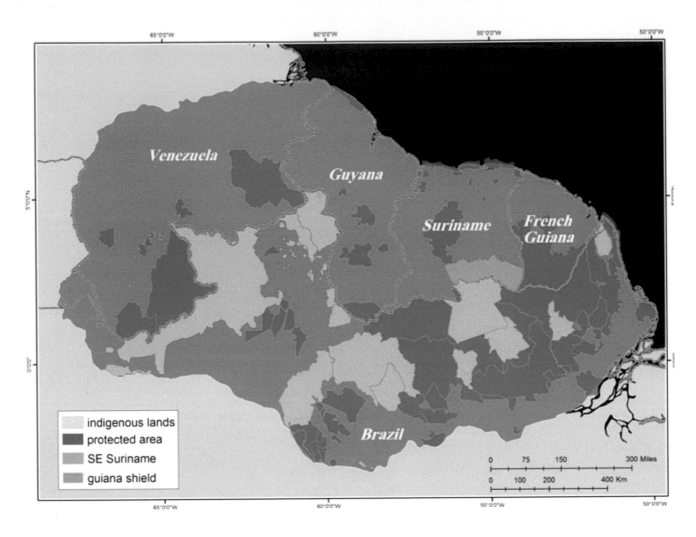

The RAP team at the Juuru Camp along the Upper Palumeu River. (P. Naskrecki)

RAP enthusiasts, from left to right: Trond Larsen (RAP Director), Fabian Oberfeld (RAP supporter), Jeffrey Kapor (RAP supporter), Russ Mittermeier (CI President), Burton Lim (mammalogist), Richard Sneider (RAP supporter), Fred Boltz (formerly CI, now Rockefeller Foundation), Richard Conniff (writer and journalist)

The mountains and extensive intact forests of Southeastern Suriname are often shrouded in clouds, and it is one of the wettest areas of the country. (T. Larsen)

Fabian (game warden), Semmie (tree spotter), and a local guide display several large anyumara fish (*Hoplias aimara*). This is an important food species for local people, and was abundant in the Upper Palumeu where fishing pressure is low. (R. Mittermeier)

The granite outcroppings and mountains of Southern Suriname encompass the headwaters that provide clean freshwater throughout much of Suriname. (R. Mittermeier)

Andrew Short collects several water beetle species new to science on top of a granite rock in the Grensgebergte Mountains. This site contained many species new to science, as well as new and unusual records for Suriname. (T. Larsen)

Sunset over the Palumeu River. (T. Larsen)

The flora and fauna of the Grensgebergte Mountains was very distinct from the surrounding lowlands. Here, a *Clusia grandiflora* fruit lies on a moss-covered liana. (T. Larsen)

The unusually strong and long-lasting rains caused the Upper Palumeu River to flood its banks, completely inundating our camp, and forcing us to move sooner than planned. (T. Larsen)

The lowland forest near our first basecamp, Juuru Camp, was distinct from the mountains. Understory palms, such as *Geonoma baculifera* (shown here), dominate many parts of the forest, especially in areas prone to flooding. (T. Larsen)

As we hastily packed up our flooded camp, we were forced to travel in boats along the same trails in the forest which we had previously used to hike! (T. Larsen)

As the rains intensified during our visit, *Spondias mombin* fruits are swept across the rapidly flooding forest floor, successfully dispersing their seeds to new locations to grow. (T. Larsen)

Water is a precious resource in Suriname, providing many of the goods and services people throughout the country depend upon, including fish. (B. O'Shea)

A storm moves over the rapids of the Palumeu River, by the small settlement of Kampu, across from our Kasikasima basecamp. (T. Larsen)

A waterfall near the base of Kasikasima Mountain. (T. Larsen)

View from an outlook of Kasikasima into the lowlands. (T. Larsen)

A Rapid Biological Assessment of the Upper Palumeu River Watershed (Grensgebergte and Kasikasima) of Southeastern Suriname

SPECIES POTENTIALLY NEW TO SCIENCE

This species of South American darter (*Characidium* sp.) from the Upper Palumeu River may also be new to science. (S. Raredon)

The RAP team discovered over fifty species that are potentially new to science. This photo shows a new genus of water beetle, discovered on granite seeps. (A. Short)

A beautiful characin fish with bright red fins (*Bryconops* sp.) that is probably new to science. (P. Naskrecki)

Canthidium cf. *minimum* is a tiny species of dung beetle (<3 mm long) collected in flight intercept traps during this survey that is potentially new to science and may belong in a new genus. (T. Larsen)

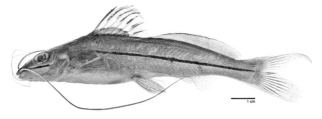

This three-barbeled catfish species (*Pimelodella* sp.) collected during the survey is potentially new to science. (S. Raredon)

This interesting catfish species (*Paratocinclus* sp.), which is potentially new to science, was collected in a tributary of the Upper Palumeu River. It has an unusual pigmentation pattern, similar to *Microglanis* catfish. (S. Raredon)

ADDITIONAL SPECIES POTENTIALLY NEW TO SCIENCE

The unusual dorsal coloration of this poison dart frog (*Anomaloglossus* sp.) differs from *Anomaloglossus baeobatrachus* found at the same sites. It may represent a species new to science, although ongoing molecular work will need to confirm this. (T. Larsen)

This tetra species (top) which is potentially new to science (*Hyphessobrycon* sp. (heterorhabdus group)) lacks the pigmentation seen in the closely related *Hyphessobrycon heterorhabdus* (bottom), both of which were collected on this survey. (P. Naskrecki)

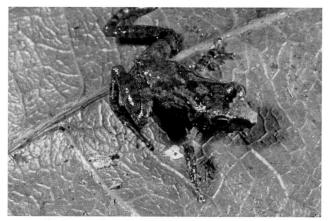

Hemigrammus aff. *ocellifer* is another attractive species of tetra from the survey that may be new to science. It is similar to the head-and-taillight tetra, which is popular in the aquarium trade. (T. Larsen)

This leaf litter frog (*Pristimantis* sp.) may be new to science, and is a member of the most diverse genus of vertebrates in the world. (S. Nielsen)

This sleek, chocolate-colored tree frog (*Hypsiboas* sp.) may be new to science. (P. Naskrecki)

This species of snouted tree frog (*Scinax* sp.) may also be new to science. (S. Nielsen)

A genus and species of katydid new to science (Pseudophyllinae: Homalaspidiini n. gen. & n. sp.). (P. Naskrecki)

An undescribed katydid species, *Artiotonus* sp. n. (P. Naskrecki)

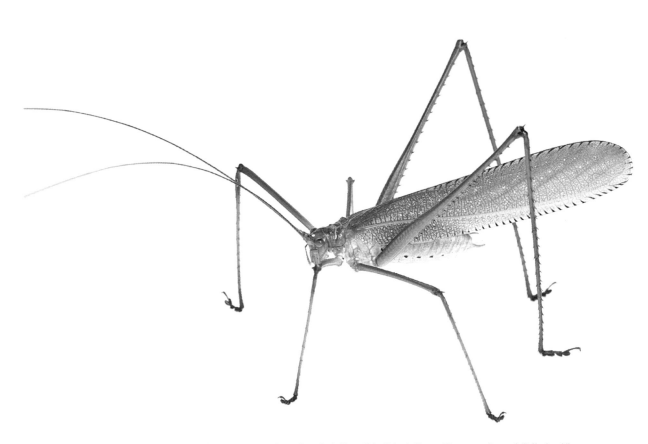

Another genus and species of katydid new to science (Pseudophyllinae: Teleutini: cf. *Macrochiton* n. gen & n. sp.) (P. Naskrecki)

ADDITIONAL SPECIES FROM THE RAP SURVEY

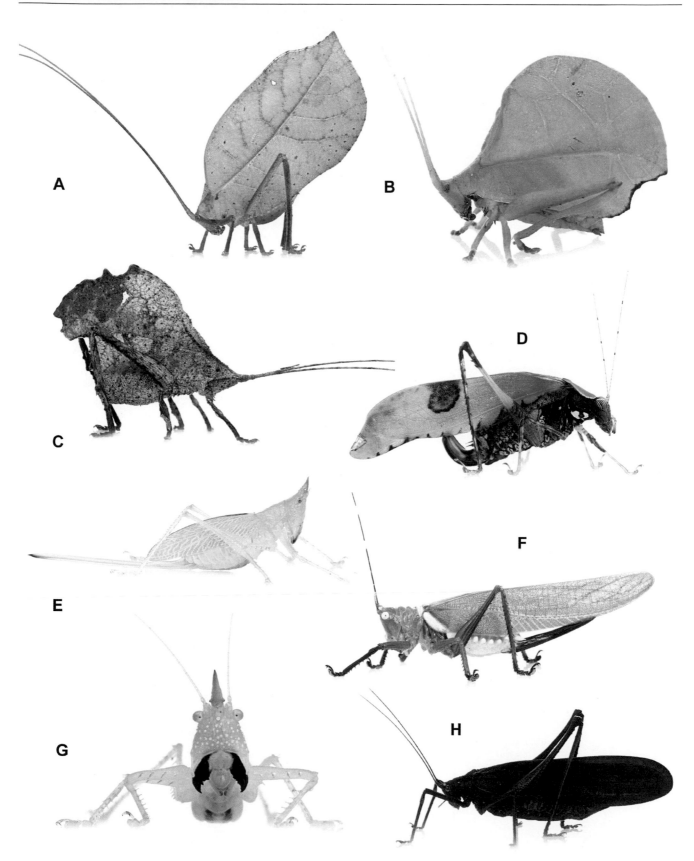

Representatives of katydid species recorded during the survey: (A) *Cycloptera speculata*; (B) *Roxelana crassicornis*; (C) *Typophyllum* sp. 1; (D) *Hetaira smaragdina*; (E) *Copiphora* sp. n.; (F) *Moncheca bisulca*; (G) *Copiphora longicauda*; (H) *Chondrosternum triste*. *Copiphora* sp. n. is a species new to science. (P. Naskrecki)

This orchid (*Epidendrum nocturnum*) was common on the granite outcroppings of Grensgebergte and Kasikasima. (P. Naskrecki)

Clusia flavida, a new tree species record for Suriname, found in the Grensgebergte Mountains. (C. Bhihki)

Another orchid found on top of the Grensgebergte mountain (*Phragmipedium lindleyanum*) is rare in Suriname. (T. Larsen)

This herb species with showy red flowers (*Costus lanceolatus* subsp. *pulchriflorus*) appears rare and localized in Suriname, and specimens from this survey represent only the fourth collection for Suriname at the National Herbarium of the Netherlands. (T. Larsen)

This survey provides the first species record for *Hirtella duckei* in Suriname. Tiny ants (*Allomerus* sp.) live in symbiosis with this plant. The plant provides shelter and food, while the ants protect the plant from herbivores. (T. Larsen)

The purpleheart tree *(Peltogyne venosa)* forms massive buttress roots which provide support. (T. Larsen)

Oxysternon festivum is a brightly colored diurnal dung beetle. These large, powerful beetles are important for burying dung, which helps to disperse seeds and regulate parasites. Males use their horn to battle with other males over mates. (T. Larsen)

Juvenile hopper. (T. Larsen)

Ants are important scavengers, and can be seen here (*Camponotus* sp.) eating a dead insect. (T. Larsen)

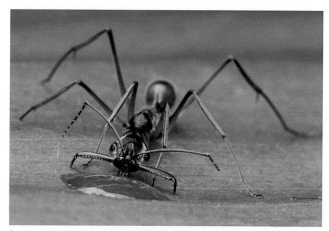

Trap-jaw ants (*Odontomachus hastatus*) swiftly snap their mandibles shut to capture prey. They have the fastest moving predatory appendages of any animal. (T. Larsen)

Rhinatrema bivitatta is a caecilian, an unusual group of primitive, limbless amphibians, most of which live underground and are rarely encountered. (S. Nielsen)

This dwarf gecko (*Gonatodes annularis*) lives in moist primary forest and can be considered a good indicator of forest health. (S. Nielsen)

The worm lizard (*Amphisbaena vanzolinii*) is another rarely encountered species, probably due to its underground lifestyle. (S. Nielsen)

Neusticurus bicarinatus is a semi-aquatic lizard found in small pools and streams in the area, and is an excellent underwater swimmer. (S. Nielsen)

The bright colors of the false coral snake (*Erythrolamprus aesculpi*) lend it protection from predators, even though it lacks the deadly venom of the true coral snake. (P. Naskrecki)

The charismatic tiger leg monkey frog (*Phyllomedusa tomopterna*) is indicative of pristine forest habitat. (S. Nielsen)

Dendrobates tinctorius is a very large poison dart frog, and different populations exhibit highly distinct color patterns. (T. Larsen)

Amereega trivitatta is a poison dart frog that advertises its toxicity with bright coloration. (T. Larsen)

A pair of tree toads (*Amazophrynella minuta*) in amplexus. As in other frogs and toads, the eggs are fertilized externally. (T. Larsen)

However, the defense of *Amereega trivitatta* is not always successful. Here, a wolf spider feeds on a young individual. (T. Larsen)

Ornithologist Serano Ramcharan holds a juvenile Dusky Parrot (*Pionus fuscus*). The RAP team rescued the juvenile from the river where it had fallen in. After a thorough drying and a few meals in camp, the parrot was able to fly back to the treetops on its own. (T. Larsen)

The Black-headed Antbird (*Percnostola rufifrons*) that occurs along the Guiana Shield represents a taxon endemic to the region. (B. O'Shea)

The White-fronted Manakin (*Lepidothrix serena*) is endemic to the Guiana Shield. (B. O'Shea)

Camera traps photographed a pair of Gray-winged Trumpeters (*Psophia crepitans*). They are social birds that eat insects and fruit from the forest floor, but also hunt snakes. (K. Gajapersad)

The Sooty-capped Hermit (*Phaethornis augusti*) was relatively common in savanna forest and open scrub on top of the Grensgebergte Mountain, but represents an important new record, since this species was previously known in Suriname only from the Sipaliwini savanna. (B. O'Shea)

The White-throated Round-eared Bat (*Lophostoma silvicolum*) is an unusual bat with enormous ears. The species is known to roost inside termite nests. (B. Lim)

The water rat (*Nectomys rattus*) is an unusual and uncommon species which we found only on the Grensgebergte mountaintop. This is a water-adapted species with webbing on its hindfeet. All individuals were caught around ephemeral water sources. (T. Larsen)

The Larger Fruit-eating Bat (*Artibeus planirostris*) was the most abundant bat during the survey. With their sharp teeth, they are capable of grabbing and eating large fruits. (B. Lim)

The bird team spotted this margay (*Leopardus wiedii*) right before dawn, perched on a nearby tree branch. This small, rarely seen cat is one of only two species in the world capable of climbing head-first down trees (the other being the Clouded Leopard). (B. O'Shea)

The Delicate Slender Opossum (*Marmosops parvidens*) is an arboreal species indicative of primary forests, and eats insects and fruit. (P. Naskrecki)

A margay curiously inspects a camera trap. (K. Gajapersad)

An Ocelot (*Leopardus pardalis*), photographed by a camera trap near the caves at the base of Kasikasima. (K. Gajapersad)

Camera traps photographed two species of deer, the Red Brocket Deer (*Mazama americana*) (shown here) and Grey Brocket Deer (*Mazama gouazoubira*). Deer are an important source of food for indigenous people. (K. Gajapersad)

The Puma (*Puma concolor*), photographed by a camera trap, is a large cat, capable of eating prey such as deer. (K. Gajapersad)

Two species of armadillo, the Great Long-nosed Armadillo (*Dasypus kappleri*) (shown here) and the Nine-banded Armadillo (*Dasypus novemcinctus*) were photographed by camera traps. (K. Gajapersad)

The Red-rumped Agouti (*Dasyprocta leporina*) is a large rodent that is common in the area, and is important for dispersing large seeds, which it caches underground and often forgets. (K. Gajapersad)

HELË KUPEPSIK LËKEN EKALËIMË TOP ËTÏ KOM UPI TOPONPË PËK

8 poinë 29 mart 2012 pona

ËITE NAPSIK RAP PËKËN KOM ËHALË TOPONPË

RAP tom kuntëm Palumë uputpëme ënïkjamkom menkatohme. Tunak watalïtom, ituponokom, wewetom malalë tëpëlamo. Mëlë katïp kunemaminum tot lo hahkatohpo Kalaiwalon ekatau.
Sin lo man akëlephak malalë man imelekalale.
Mële Surinam lon.
Sin lo po man 16 me tïwe tïwe ituhtau eitop ëhekatïpïla.
Tïwëlen po tïwe wewetom, tïwëlën po tïwe sihnahtom malalë ilomonme ïpi tom kawë mëisnë la lome tïwëlenpo kawë 780 meiter ïpi tom.

RAP tom kuntëm 4 me patatom pona
1. Palumë uputpëme tuna amahtom apsik lome kopëme aptau tïhnëlë tupke tëtïhe malalë ituhtailë tikunai asimhak mëje tuna etpitailë lo kawëkawë ituhtalë ïpïtom tëpuhpe malalë ituhtailë lo pulapulam malalë tëputom kahwënotompëk ituhpe mëje helë ipohnëptop ametajeinë mïtëjai aktuhpona aptau kawëna katïp mïtëjai mëlëkatïp man mëje eitop Palumë uputpëme 270 meiter kawë lo eitop tunapeptapëk.
2. Helë lo hahkatop lotïp ïpï ehet Meutëhpa helë kawënma tïmaikiptëi ejahe ïpï 800 meiter kawë iweitop malalë tënei ejahe ïpïpo kawë tëputom wewetom ekutom enepophak kankom tënei ejahe helëpo ëtïkompalëmalë ituhtau awoname kolela itu awonamehapon lëken ituhtau eitop malalë ïpïtom awoname pajahpelëken kopïnme malalë lo hamunme lëken ituhtau eitop helë.
3. Makulutu amat man Palumë uputpïme Palumë tuna ëtamatsahkatop helë Makulutu amat, malalë kopë me aptau tïnëlënma tïkunai ituhtailë.
4. Palumë tuna ailë man tëpu ehet Kasikasimë tëpu man pëhkënatpë lëken lome man helë tëpu peptame malalë kawë iweitop 700 meiter uhpolo. Helë lome helë katïp

lëken tïmaikiptëi ejahe uhpolep kawë iweitop.kawë lo eitop 200 meiter tuna pepte pëk kawë ïpïpo tëweihamo wewetom ëmëmïla tupke tuna aptau kasikasimë tëpupëk awoname ëtïkompalëhpe ipëk ekutom enepopkakankom wewehpe huwa

HELË TOHME RAP TËMAMINE EITOPON PË

Surinam lon po tëweihem helë lo akuwinpï itumehlëken tëweihem helë lo po.
Helë lo man tïkulunmaihe akuwinpï me iwesike, wewetom, tunatom. Helë man iloptailë Surinam pono kom moja malalë ipajamonpï kom moja wantë aptau. Malalë man iloptailëhnë Surinam lon ekatalï tom moja. Iloptailon helë tuna lo itu huwa helëkom ohanë matop ipokela. Masike helë itu lo malë man tïpïne talankom man helë itu tuwalëla masike mantot helë ituhtalïtom malë imenkahe ëhmelë enehe tuwalë eitohme.
RAP toman ëtïkom palë enehe man tot ëhmelë mënimenketot tamanu sipïnanïpjatëu itu ëhmelë kulonkom tïkai.

HELË TUPITPÏ TOM MALALË ENETPÏ TOM EKALË TOP

Sin enelihtau man tënei ejahe kolepsik imelekapïntom ëtikom kaliponotom ënimelekalale eitop.
Malalë man tënei ejahe mële ituhtau ënïkjamkom tïwëlenpo imna eitop lome molo lëken ispe.
Malalë man tuna molo ipok mëisnë masike tïwetïwelën katom mohpe. Malalë ipokan tuna malalë iloptailë tëweihamo helë tuna tom.
Malalë imelekalale eitop tom helë kaliponotom ënimelekalale eitop. Mëlë
katïp man enehan mëkja nenetpï. Malalë man ispe 50 uhpolo tïwetïwëlënkom enelale eitop masike ihjan.
Mëlë katïhman katom 11 me tënei ejahe, 6 me pëlëtom malë , 1me ëkëi. Malalë kolenma akinme mïphak hapon tënei ejahe tïwe tïwëlën tëweihamo.
Masike RAP tom mënke helë kom man iloptailë tuna tom malalë itu tom Surinam lon po.

HELË TA ITU PÏNANÏP TOP

Malalë man sin po ituhtau eitop ipok enepophak huwa.
Malalë kolelan ma itumehle tëweihem.
Malalë sin lo po pëtuku wewetom malalë tëpëlamo malalë ekutom. Tïwelën kom lo po imna eitop malalë Guyanapata tom pohnë imna malalë ëhmelë lo kuptë imna eitop.
Malalë man Suriname lon po tuna tom.
Malalë man mëlë ipok tuna tëlïhem me malalë tëhamo masike man iloptailë Surinam lo po tëweihamo ja helë tom.
Malalë man mëlë kom iloptailë tëpïmahamo katohme malalë man. Helë tuna , ipok wapot stroom ïtohme tapuhe aptau tïkaitot , helë tunapëk malë helë ipohnëptop lëken ëjahe tuwalë ëweitopkome kohle ëtïkompalë ïtohme helë tuna

ipok tïkaitot masike ënawohanë mapola eitëk ipokanmëisnë helë tuna tïkaitot ,
malalë man itu iloptailë tohme uwa helëkom man ipok tëhamoja malalë man iloptailë kaliponotom moja tohme uwa epitom man ituhtau malalë wewetom pakolotom ïtohme kaliponotom moja.
Malalë man itu iloptailë ëhmele tïwelen loponokom moja masike mënekaleja tupina nom helë iloptailë eitohpëk tohme tïwëlen po man tuna hautuhpe lome mëlë po tuna ipok masike man iloptailë helë.
Masike helë pëtuku kutïtei tuna helëlë katïp eitohme ipoklëken tuna eitohme helëpo malalë man iloptailë mïhja hemalëlëken nula mïhja wantë eitop ponaman helë iloptailë.

Helë ëtïkom palë malalë mëkpalëtom miliktoponpë RAP pëkën kom temamine aptau mart 2012 po

Ëhmelë ituhtalï tom pëk mëk palë tom	Kampu 1po Palume uhpoinë 277 m	Kampu 2 po Meutëhpa 790–820 m	Kampu 3po kampu Makulutu amat kumtalï 240–260 m	Kampu 4 po kasikasimë 201 m	Ëhmelë kampu 3 me eitop helë	Mëham ihjan tom enetpïtom mëham pon imna tïwëlëpo lome tënei talë kalipono tom ënenelale eitop ituhtalïtom	Mëham akenameh le inenet pïkom Surinam lon po
Tëpepelamo tom	161 me	68 me	27 me	161 me	354 me		15 me
Simalaimë mëhe tunakwalï	92 me	10 me	-	105 me	157 me	26 me	Sinkomohnë man upilahnë
Kataputpë tom	93 me	40 me	-	74 me	107 me	10 me	Sinkomohnë man imelekalahnë
Ëlukë jum tom	29 me	2 me	-	34 me	52 me	6 me	
Mïphaktom	72 me	25 me	-	92 me	149 me	Sinkomohnë man imlekalahnë	Sinkomohnë man imelekalahnë
Katom	71 me	-	16 me	49 me	94 me	11 me	2–4 me
Lopo tëhatëhalë hamo ololi haponkom ëkëihaponkom huwa	26 me	7 me	-	21 me	42 me	1 me	
Pëlë tom	30 me	6 me	-	24 me	47 me	6 me	
Tolopïh tom tïwëtïwë lënkom	196 me	103 me	-	233 me	313 me		
Ituhtalï tom jakin tom tïpei nomtahamo ëhmelë	25 me	12 me	-	23 me	39 me		
Ituhtalïtom junutpëtom ëhmelë tïpeinomtahamo pëk helë	16 me	-	-	20 me	24 me		
Ëhmelë	811 me	273 me	43 me	836 me	1378 me	60 me	17–19 me

Helë imilikutpï tom tïpïne tëweihamo RAP tëmamine aptao enetpïtom 2012 aptau

Tïkena	Tënonkom	CITES sin IUCN katop Tïkena
Tëpelamotom		
	Vouacapoua americana	CR
	Syagus stratincola	VU
Orchidaceae	*Cleistes rosea*	CITES Appendix II
Orchidaceae	*Dichaea picta*	CITES Appendix II
Orchidaceae	*Epidendrum densiflorum*	CITES Appendix II
Orchidaceae	*Epidendrum nocturnum*	CITES Appendix II
Orchidaceae	*Maxillaria discolor*	CITES Appendix II
Orchidaceae	*Phragmipedium lindleyanum*	CITES Appendix II
Tëpëlamotom		
	Pipile cumanensis	VU
	Crax alector	VU
	Myrmotherula surinamensis	VU
	Patagioenas subvinacea	VU
Ituhtalïtom junutpëtom ëhmelë tïpeinomtahamo pëk helë		
	Pteronura brasiliensis	EN
	Tapirus terrestris	VU
	Ateles paniscus	VU

Mëkpalë tom tïkena kolela tëwei hamo: VU=Anumna tëwei hamo , CR=Kolelanma mantot (Enek www.redlist.org) malalë (enek www.cites.org)

Kort Overzicht van het Rapport

DATA VAN DE RAP-STUDIE

8–29 maart 2012

BESCHRIJVING VAN DE LOCATIES VAN DE RAP-STUDIE

Het RAP-team heeft de aquatische en terrestrische flora en fauna van het stroomgebied van de Boven-Palumeu Rivier in zuidoostelijk Suriname, nabij de grens met Brazilië bestudeerd. Dit gebied is mogelijk het meest verafgelegen, ongerept en onverkend regenwoudgebied van Suriname. Het gebied is volledig bebost met 16 verschillende landschapssoorten die gaan van laaglandonder water gelopen bos tot geïsoleerde bergtoppen hoger dan 780 m. Het RAP-team heeft vier locaties bestudeerd: 1) de Boven-Palumeu Rivier, waar de rivier een kleine kreek wordt, vooral met seizoensgebondenonder water gelopen bos met enkele hoog drooglandbossen op granieten heuvels en een zwampbos op ~270 m boven zeeniveau; 2) Grensgebergte bergtop op 800 m, open granietrots omringt door bos, met een mengeling aan vegetatiesoorten, waaronder cypergrassen en bromelia's met orchideeën en *Gesneriaceae* op de hellingen, lage struikvegetatie op de rotsen, laag savannebos en drooglandbos op granieten heuvels; 3) Makrutu Kreek bij de samenvloeiing met de Boven-Palumeu Rivier en de Makrutu Kreek, aquatische ecosystemen en seizoensgebonden bos langs de waterwegen; en 4) de Midden-Palumeu Rivier en Kasikasima Gebergte, een unieke granieten bergketen die meer dan 700 m boven het regenwoud uittorent met vegetatie die vergelijkbaar is met die van het Grensgebergte, en laagland (~200 m boven zeeniveau), seizoensgebonden onder water gelopen bos, hoog drooglandbos op granieten heuvels en savannebos.

REDENEN VOOR EEN RAP-STUDIE

Zuidoostelijk Suriname is een van de laatste uitgebreide, ongerepte regenwoudgebieden op aarde. Behoud van deze bossen en rivieren en de natuurlijke diensten die zij verlenen aan het volk van Suriname zijn belangrijk voor de toekomst van het land en de regio. Aangezien bijna niets bekend is op wetenschappelijk gebied, is de eerste stap in het beschermen van zuidoostelijk Suriname het verzamelen van basis biologische en sociaal-economische data voor de regio. Deze RAP-studie levert data op voor het sturen van behoud en duurzame ontwikkelingsactiviteiten in zuidoostelijk Suriname en voor het verschaffen van de wetenschappelijke rechtvaardiging voor de bescherming van deze gevarieerde en belangrijke regio.

BELANGRIJKSTE RESULTATEN

Resultaten van alle taxonomische groepen die zijn bestudeerd tijdens de RAP-studie laten zien dat zuidoostelijk Suriname een zeer hoge biodiversiteit bevat en zich in een maagdelijke toestand bevindt met bijna geen menselijke invloed. Alle taxonomische groepen, met uitzondering van de grote zoogdieren wijzen erop dat zuidoostelijk Suriname uniek is in vergelijking met andere gebieden van het Guianaschild, dat veel soorten bevat die nergens anders worden gevonden. De samenstelling van de plantensoorten verschilt van de flora van noordelijk Suriname en er komen verschillende vogelsoorten voor in zuidoostelijk Suriname die niet in het noorden worden aangetroffen. De waterkwaliteit en visdiversiteit zijn hoog, wat betekent dat het gebied, waar de bronrivieren van veel van de belangrijkste rivieren in Suriname ontspringen, overvloedige zoetwaterbronnen bevat. De variatie in hoogteverschillen binnen de gebergteketens en de ongerepte natuur van het laagland bos in zuidoostelijk Suriname draagt bij aan de hoge biologische diversiteit in de regio. We hebben meer dan vijftig soorten gevonden die waarschijnlijk nieuw zijn voor de wetenschap, waaronder elf vissen, zes kikkers, een slang en vele insecten. De RAP-resultaten tonen het belang van de diversiteit van de bossen, soorten en stroomgebieden van zuidoostelijk Suriname aan.

Soorten vastgelegd tijdens de RAP-studie in zuidoostelijk Suriname, maart 2012

Taxon	Kamp 1 Boven-Palumeu (Juuru) 277 m	Kamp 2 Grensgebergte 790–820 m	Kamp 3 Makrutu Rivierkamp 240–260 m	Kamp 4 Kasikasima 201 m	Totaal (alle 3 kampen)	# soorten mogelijk nieuw voor de wetenschap	# soorten voor het eerst gerapporteerd voor Suriname
Planten	161	68	27	161	354		15
Aquatische kevers	92	10	-	105	157	26	NTB*
Mestkevers	93	40	-	74	107	10	NTB
Sabelsprinkhanen	29	2	-	34	52	6	
Mieren	72	25	-	92	149	NTB	NTB
Vissen	71	-	16	49	94	11	2–4
Reptielen	26	7	-	21	42	1	
Amfibieën	30	6	-	24	47	6	
Vogels	196	103	-	233	313		
Kleine zoogdieren	25	12	-	23	39		
Grote zoogdieren	16	-	-	20	24		
Totaal	811	273	43	836	1378	60	17–19

*NTB = Nog te bepalen

Soorten van belang voor natuurbehoud gedocumenteerd tijdens de 2012 RAP-studie in zuidoostelijk Suriname

Groep	Soorten	IUCN of CITES categorie
Planten		
	Vouacapoua americana	CR
	Syagus stratincola	VU
Orchidaceae	*Cleistes rosea*	CITES Appendix II
Orchidaceae	*Dichaea picta*	CITES Appendix II
Orchidaceae	*Epidendrum densiflorum*	CITES Appendix II
Orchidaceae	*Epidendrum nocturnum*	CITES Appendix II
Orchidaceae	*Maxillaria discolor*	CITES Appendix II
Orchidaceae	*Phragmipedium lindleyanum*	CITES Appendix II
Vogels		
	Pipile cumanensis	VU
	Crax alector	VU
	Myrmotherula surinamensis	VU
	Patagioenas subvinacea	VU
Zoogdieren		
	Pteronura brasiliensis	EN
	Tapirus terrestris	VU
	Ateles paniscus	VU

Rode Lijst van de IUCN van Bedreigde Diersoorten categorieën: VU=Kwetsbaar, CR=Ernstig bedreigd (zie www.redlist.org); CITES (Convention on International Trade of Endangered Species and Wild Flora and Fauna - Overeenkomst inzake de internationale handel in bedreigde soorten wilde dieren en planten) Appendices I, II en III (zie www.cites.org)

AANBEVELINGEN VOOR NATUURBEHOUD

Zuidoostelijk Suriname is een mondiale natuurschat. Er zijn slechts weinig plaatsen op aarde die nog zo maagdelijk en ongerept zijn als dit gebied. De regio bevat een hoge diversiteit aan planten en dieren en een unieke samenstelling van soorten die het gebied onderscheidt van andere delen van Suriname, het Guianaschild en de wereld. Zuidoostelijk Suriname bevat de bronrivieren van een aantal van de grootste rivieren van Suriname die in schoon drinkwater voorzien en voedselbronnen voor het volk van Suriname. Ook ondersteunen zij de landbouw- en energieproductie benedenstrooms. De ongerepte bossen van het gebied voorzien in een continue voedselbron, medicijnen en bouwmaterialen voor de lokale bevolking. Op regionaal en mondiaal niveau helpen zij het klimaat te reguleren. Onze bevindingen laten ook zien dat zuidoostelijk Suriname van disproportioneel belang zal zijn om te voorzien in zoetwaterbronnen onder toekomstige klimaatveranderingsscenarios. Wij bevelen sterk de bescherming van zuidoostelijk Suriname aan voor het behoud van de unieke en diverse soorten en de zoetwaterbronnen voor het volk van Suriname en de wereld, nu en voor toekomstige generaties.

Executive Summary

INTRODUCTION

Suriname is one of the last places on Earth where an opportunity still exists to conserve huge tracts of pristine diverse tropical forests. Suriname is less than 6% deforested, exhibits the lowest population density of any moist tropical region on Earth (0.2 people/ha), has few roads in the forested part of the country (which can be accessed only by small boat, small plane, or on foot), and virtually all of the lands are public and under the control of national government or indigenous and Maroon communities.

However, the isolation that has protected Suriname's ecosystems, natural resources, and indigenous cultures is disappearing at an increasing rate, and the opportunity to act to preserve these remarkable resources will soon be gone. Record high commodity prices have encouraged the spread of illegal gold miners from Brazil across the region, spurred potential major hydropower and mining investments, and provided the incentive to press ahead with road and dam projects.

Southeastern Suriname is possibly the most remote and pristine region of Suriname. The region extends from the Sipaliwini Nature Reserve in the west, across the mountain ranges of the Grensgebergte, Toemoekhoemak gebergte, and Orangebergte, to the border with French Guiana. It is bordered to the north by the Tapanahony River, which becomes part of the Marowijne River watershed. Southern Suriname has a rich biodiversity, making it a high priority region for protection. This region was highlighted in the Guiana Shield Priority Setting Workshop held by CI, IUCN and UNDP in April 2002 (Huber and Foster 2003) as one of the highest ranked areas for biodiversity conservation. One of the first steps in the process of protecting Southern Suriname is to collect baseline biological and socio-economic data for the region, particularly in areas where no scientific data exist.

Conservation International-Suriname and CI's RAP program began this process by collecting biodiversity data in August 2010 in Southwestern Suriname near the Amerindian village of Kwamalasamutu (see results in O'Shea et al. 2011). The 2012 RAP survey of Southeastern Suriname was the next step in establishing a baseline of data for Southern

Suriname. Together, these RAP surveys provide data to guide conservation and development activities in Southern Suriname and to provide scientific justification for protection of this diverse and important ecosystem.

The Grensgebergte and Kasikasima Mountains and Palumeu River

The RAP survey provided scientists with the rare opportunity to explore a totally unstudied and unique mountain range, as well as the Upper Palumeu River. To our knowledge, the mountains of the Grensgebergte (Border Mountains) and Upper Palumeu River have never been studied or even explored by scientists. The local Amerindians occasionally travel up the Palumeu River and walk over the border to Brazil but few ever enter the Grensgebergte Mountains. The elevation and forest types within the Southeastern Suriname mountains range from lowland floodplain forest to isolated mountain peaks at over 780 m elevation. The Palumeu River flows in a wide floodplain within the Grensgebergte but the river is shallow and difficult to navigate by boat. Thus the only way to enter the mountains is to cut a new trail and enter on foot, or to enter by air. The first step for this RAP survey was to identify rocky outcrops in the mountains from satellite imagery where a helicopter could possibly land. The Kasikasima Mountain is a unique granitic mountain formation that rises over 700 m above the rainforest. It has over twelve peaks of steep granite outcroppings. While the METS tourism company takes adventurous tourists to the rock, there have not been many studies of the biodiversity of the area. The mountain has a trail that allows access up to about 500 m within the forested side of the mountain.

The RAP Team

The RAP scientific team of 16 people included scientists from the Anton de Kom University of Suriname, CELOS, Conservation International, Global Wildlife Conservation, the Museum of Comparative Zoology at Harvard University, the North Carolina Museum of Natural Sciences, the Biodiversity Institute at the University of Kansas, the Royal Ontario Museum, and the National Herbarium of the Netherlands—Naturalis Biodiversity Centre. The RAP team collected data on water quality, plants, and the following groups of animals: aquatic beetles, dung beetles, ants, katydids, fishes, reptiles and amphibians, birds, and small and large mammals. The RAP team was accompanied by many local AmerIndian guides and field assistants, a film crew from Media Vision, a writer and a photographer sponsored by the National Geographic Society, two game wardens from the Nature Conservation Division, two medics, and staff from Conservation International-Suriname.

CI's Rapid Assessment Program (RAP)

RAP is an innovative biological inventory program designed to use scientific information to catalyze conservation action. RAP methods are designed to rapidly assess the biodiversity of highly diverse areas and to train local scientists in biodiversity survey techniques. Since 1990, RAP's teams of expert and host-country scientists have conducted over 80 terrestrial, freshwater aquatic and marine rapid biodiversity surveys and have contributed to building local scientific capacity for scientists in over 30 countries (Alonso et al. 2011). Biological information from RAP surveys has resulted in the protection of thousands of hectares of tropical forest, including the declaration of protected areas and the identification of biodiversity priorities in numerous countries.

The Grensgebergte and Kasikasima RAP Adventure

The Grensgebergte and Kasikasima RAP survey took place from March 8–29, 2012. Prior to these dates, an advance team of 21 men from the Amerindian villages of Apetina (Wayana) and Palumeu (Trio and Wayana) led a reconnaissance mission to locate and set up the first RAP base camp (RAP Site 1) from February 26–March 7, 2012. They were accompanied by Krisna Gajapersad (RAP coordinator from CI-Suriname), three CI field workers (Andre Semmie, Hermando Banda, and Jeffrey Krimbo), along with Ted Jantz and Rafael Jantz from MediaVision, who filmed the adventure. The team traversed and carried six heavy boats around many treacherous rapids to get far up the Palumeu River. The water level was very high, which made the trip all the more dangerous and lengthy. They reached Juuru camp on March 6, a site where the Trio and Wayana occasionally camp on their way to Brazil. From this site, a trail to the border with Brazil commences. The men from Apetina quickly got to work to clear a 30 m × 30 m site within the forest within which a small helicopter could land. They chose the site of a former small agricultural plot so cutting was a bit easier. They also set up a very large base camp consisting of two large tents (tarps over poles) for hanging hammocks in which to sleep, a kitchen tent, an eating tent, and a laboratory tent for the soon to be arriving scientific team. They also built several smaller tents to sleep in themselves. The camp was at the bottom of a steep hill, right along the Palumeu River, which at this point was a fairly small creek.

The RAP scientific team of 16 scientists, accompanied by a journalist and a photographer from National Geographic Magazine, two game wardens from Nature Conservation

Division, and a medic flew from Paramaribo to Palumeu Village on March 8. They brought along with them about 2000 kilos in food and scientific equipment! Some of this was stored by METS in Palumeu while most went to the base camp with them. The RAP team spent one night in Palumeu at the METS camp. On March 9, a helicopter from HiJet arrived in Palumeu around 10 am (to the delight and wonder of the people of Palumeu!). The RAP team and gear were transported from Palumeu Village to the RAP base camp (RAP Site 1) over the following three days. Each flight to base camp took approximately one hour round trip. On one trip, the helicopter had mechanical difficulties while at the base camp and was almost stuck there overnight. However, the helicopter mechanic was fortunately along on the ride and was able to fix the problem well enough for the helicopter to return to Palumeu. However, a problem with the rotor motor was found and a part had to be ordered from Miami. In the meantime, HiJet sent a replacement helicopter to assist the RAP team. The rest of the RAP team and equipment finally arrived at RAP base camp on March 11 in the new helicopter.

From this base camp, the RAP team sampled in the lowland forest nearby and in the Upper Palumeu River (RAP Site 1) and also flew by helicopter to a mountain top in the Grensgebergte Mountains (RAP Site 2) with an open rocky outcrop and a flat area upon which the helicopter was able to land. The RAP team visited this mountain top in small numbers (2–6 people per day) with some people staying overnight or a few nights on the mountain. Fortunately, the weather was clear enough for helicopter flying and landing on the mountain most days. Only on March 17 was the weather too cloudy for the helicopter to land in the morning, but it was able to pick up stranded RAP team members in the late afternoon. There must have been a lot of rain in the upper reaches of the Palumeu River that day (and the previous day) for the river started to rise and by 6 pm on March 17 had flooded most of the base camp. The trail that once connected the base camp with the helipad was now completely underwater and had to be traversed by boat (see page 20). This area was known to flood by the local Amerindians, but the RAP team was hoping that it would not flood during the week that the team spent there! However, it did and the water continued to rise during the night.

Thus on March 18, the RAP team evacuated the first base camp and set off toward the next camp at Kasikasima on the Middle Palumeu river. The botany, water quality, and fish teams, along with all the Amerindians and most of the support crew headed out in boats. They were able to sample a third site (RAP Site 3) at the confluence of the Palumeu River with the Makrutu Creek en route. Eleven of the RAP scientists along with the National Geographic team departed by helicopter and landed at Kampu, a small village across the river from the Kasikasima Mountain. The scientists spent one night there. The next day, Krisna Gajapersad (who had left the field on March 9 for a week in Paramaribo) arrived at the village by boat with Priscilla Miranda from CI-Suriname and nine men from Palumeu. The men immediately set to constructing a base camp (RAP Site 4) for the RAP team along the river on the side of the Kasikasima Mountain. They constructed two large tents for sleeping, a kitchen tent, and a laboratory tent. The RAP team was able to move in the next day. The rest of the RAP team arrived the next day after a challenging three-day voyage by boat. From this camp, the team was able to reach the Kasikasima mountain by a three hour hike along well established trail system set up by METS for tourists that they bring out to this area. The METS camp was located about one hour hike along the river from the RAP camp.

While the team was now closer to Palumeu, they still had to get around several dangerous rapids on the Palumeu River to get back. The now 29 men from Apetina and Palumeu bravely spent several days carrying and dragging the large dugout boats along the trail 3 km through the forest from the RAP camp to the METS camp to get around the largest of the rapids. They also carried food and gear this distance. The RAP team was amazed at their strength and very grateful for their support.

The RAP team departed the Kasikasima base camp in two stages so that boats could be used to transport people and gear to Palumeu and then return for more. Ten people departed on March 26 and the rest departed on March 28. Both teams were able to get to Palumeu Village after about one hour walk (around two rapids) and a four-hour boat ride.

The RAP team then spent one night in Palumeu Village to present their preliminary findings and tell their adventures to the Captain and people of Palumeu. They returned to Paramaribo on March 29 to prepare the preliminary report and to present their findings to partners and supporters in Paramaribo on March 30.

Description of RAP survey sites
Olaf Bánki

The RAP team surveyed around four main sites along the Upper and Middle Palumeu River: Juuru Camp, Grensgebergte Rock, Makrutu Creek, and Kasikasima Camp. Only the coordinates of the four base camps are given here; most sampling was done within 5–10 kilometers of these camps. Certain groups sampled in other areas as well (e.g., along rivers between camps). Please refer to individual chapters for sampling protocols and localities.

Site 1. Juuru Camp (Upper Palumeu River)
N 2.47700, W 55.62941
Elevation 277 m asl.
9–18 March 2012

The first camp was situated on the west bank of the Upper Palumeu River. This place is known as the boat landing or 'Tiyaring' (Trio) for the trail to Brazil and is used by the Amerindians of Palumeu and Apetina as a temporary camp-site (Juuru Camp). The camp was situated in 'tall seasonally flooded forest', which was unmistakingly proven when the camp flooded on the 17th and 18th of March. During our stay in this camp it rained almost every day and part of days.

The Upper Palumeu River is a meandering river similar to the Kutari River and Wioemi Creek in the vicinity of Kwamalasamutu where the 2010 RAP survey took place. The Upper Palumeu River is situated in a hilly landscape in which the valleys have been filled with erosion material (loamy and sandy substrate). The river has cut its way through the eroded sediment. In the valleys of the hilly landscape 'seasonally flooded palm swamp forest' can occur, and in the river bends downstream of Juuru Camp 'tall herbaceous swamp vegetation and swamp wood' can be found. Most aquatic sampling was done in the Upper Palumeu River, up and downstream of Juuru Camp and in its tributary creeks.

On top of the hills 'tall dryland tropical forest on laterite/granite hills' did occur which was intersected with large granite boulders. Rocky places with species typically found in open rock vegetation were also found in the forest. Most terrestrial collecting was done along a trail that went up from the camp to a hill of 417 meters above sea level, and along the trail to Brazil. Our sampling trail to Brazil passed through 'tall seasonally flooded forest', 'secondary forest', 'tall dryland tropical rain forest', 'tall swamp forest', 'bamboo forest', and ended at a large waterfall. The secondary forest patch, which was an abandoned agricultural field, was cut open to serve as a helipad.

Site 2. Grensgebergte Rock
N 2.46554, W 55.77034
Elevation 790–820 m asl.
12–18 March 2012

The second camp was a small satellite camp based on top of a granite mountain within the Grensgebergte at an elevation of 790 to 820 meters above sea level. This mountain peak was one of the highest points of the Grensgebergte observed. It could only be reached by helicopter, and was approximately 16.5 km from the helipad of base camp 1. The mountain was pre-selected by using a landscape classification based on Landsat imagery and by flying over the area in a small Cessna airplane a few months in advance of the RAP. The helicopter landed on the rock on a quite open spot with low shrubs.

The Grensgebergte is a mountain range of rolling mountains with steep granite rock faces. The name Grensgebergte literally means 'border mountains.' The mountains are covered with forest and only a few mountain peaks have places with open rocks. On the mountain a mosaic of vegetation types could be found. At the slopes we observed bare rocks with seeping water with carnivorous water plants forming ideal habitats for water beetles. The occurrence of carnivorous plants indicates the low nutrient status that some of these micro-habitats have for plants. Patches of cyper grasses and bromeliads with orchids and gesneriads could be found on the slopes as well. We observed a tortoise of 30–40 cm long in wet patches surrounded by cyper grasses and small herbs on the mountain plateau. On top of the mountain peak some areas were covered with low shrub vegetation. All these herb and shrub vegetation is generally referred to as 'open rock (Inselberg) vegetation'. We also observed 'short savannah forest on granite rock' with mosses, potentially formed by moisture from low hanging clouds. Most of the mountain peak was covered by a mixture of 'savannah forest on granite rock' and medium sized 'dryland tropical forest on granite hills'. A small flowing creek was spotted between two large ridges on lower elevation. Due to logistical constraints both the forest and the creek could not be properly assessed in terms of species composition.

Site 3. Makrutu Creek
N2.793311, W 55.367445
Elevation 240–260 m asl.
18–21 March 2012

The third camp was situated at the junction of the Upper Palumeu River and the Makrutu Creek. Only the fish, plant, and water quality RAP teams visited this camp by boat from the first camp. The first stretch of the journey the Upper Palumeu River was frequently meandering with many trees growing in the river bends. Almost the entire river was associated with 'tall seasonally flooded forest' accompanied by overhanging tree branches above the water. From the point onwards where the Tapaje Creek was flowing into the Palumeu River the meandering got less and the river banks of the Palumeu River became steeper. The steep banks were covered with 'high dryland tropical rain forest'. On less steep spots we observed 'tall swamp forest', and 'tall herbaceous swamp vegetation and swamp wood'.

Rapids occurred close to the junction with the Makrutu Creek. The camp was situated on rocks within the rapids just downstream of the junction of the Palumeu River and the Makrutu Creek. On top of these rocks a mixture could be found of 'open rock vegetation' dominated by bromeliads, 'short savannah forest on granite rock' and secondary forest. The meandering Makrutu Creek is accompanied by 'tall herbaceous swamp and swamp wood', and 'tall seasonally flooded forest'. Most collecting took place upstream of the Palumeu River from the Makrutu camp, along the Tapaje Creek, and along the Makrutu Creek.

Site 4. Kasikasima camp
N2.97731, W 55.38500
Elevation 201 m asl.
18–28 March 2012

Part of the team that did not join RAP Camp 3 were transferred from the first camp by helicopter to a small settlement called Kampu along the Palumeu River on the opposite of the river from the Kasikasima Mountains. On the 20th of March the RAP teams moved into the camp that was built on the opposite side of the Palumeu River from the settlement of Kampu. The fish, plant, and water quality RAP teams joined the group at this camp on the 21st of March.

On the journey from the Makrutu camp to the Kasikasima camp we crossed two major rapids where the boats had to be unloaded and pulled. Within these rapids large forested rocks occurred. The forest on top of these rocks can be classified as 'short savannah forest on granite rock', with some species very typical for open rock places. The rapids contain 'open rock vegetation' with seeping water and some carnivorous plants similar to the vegetation that was found on the granite mountain (site 2) on much higher elevation. Steep banks also occurred along the Palumeu River.

The camp was situated along the river where large rock formations were found in the forest and into the river creating a large relatively shallow rapid. 'Short savannah forests on granite rock' occurred along the river and on top of the hills where the parent rock reached the surface of the soil. From the river bank a landscape of undulating granite hills started instantly. These granite hills were covered with 'tall dryland tropical forest on laterite/granite hills'. Creeks, with palm swamps and some creek vegetation and forests could be found in the sharp gullies between the mountains. Large trees occurred on the slopes. Closer to the Kasikasima Mountains, the amount of large boulders in the landscape increased. 'Short (dry) savannah forest on granite rock' with palms intersected now and then by a creek, could be found on the Kasikasima Mountain. The 'short savannah forest' on top of the Kasikasima Mountain (ca. 510–790 m asl.) was dominated by mosses, bromeliads, and ferns, which indicates frequent moist conditions due to low hanging clouds. The open rock vegetation on this mountain was similar to that of the Grensgebergte, dominated with bromeliads, orchids, and gesneriads.

OVERALL RESULTS

Results from all of the taxonomic groups surveyed during the RAP survey reveal that Southeastern Suriname contains very high biodiversity with many unusual and endemic species. We found over fifty species that are probably new to science, including eleven fishes, six frogs, one snake, and many insects. Our survey also indicated that ecosystems of this region are extremely healthy, with good water quality and virtually no human impact. High abundance of large birds such as cracids and parrots and records of large cats and other mammals including primates, coupled with high dung beetle abundance and biomass, indicate that hunting pressure is low or non-existent. The occurrence of a high diversity of all taxa, including large and medium-sized mammals, indicates a pristine and productive ecosystem.

For all of the taxonomic groups except the large mammals, the RAP data indicate that Southeastern Suriname is unique from other areas of the Guiana Shield, containing many species not found elsewhere. Plant species composition differs from the flora of Northern Suriname and several bird species occur that are not found in the northern parts of the country. The flora and fauna of Southeastern Suriname is also distinct from nearby Southwestern Suriname. For example, about half of water beetle and katydid species documented in Southeastern Suriname were different from those recorded in Southeastern Suriname (Short and Kadosoe 2011). Dung beetle communities were also distinct from Southwestern Suriname. This is especially remarkable given the short distance between these two areas, as well as the fact that the lowland forests of Southern Suriname are often considered fairly homogenous. Diversity, standardized by sampling effort was also generally higher in Southeastern Suriname.

A high degree of heterogeneity was found between the three RAP survey sites within Southeastern Suriname. As may be expected, the higher elevation sites on Grensgebergte (Site 2) and Kasikasima Mountain (Site 4) have a distinct flora and fauna compared to the surrounding lowlands. Four of the 15 new plant records were found in the transition between short savanna forest on granite rock and open rock vegetation at the Grensgebergte. Five of the 15 new plant records including one new plant genus for Suriname were recorded in the hilly landscapes surrounding the Grensgebergte and the Kasikasima Mountains. The surroundings contain several dominant vegetation types that are floristically distinct for the central and southern parts of Suriname. The bird fauna of the Grensgebergte contained several species with ranges restricted to higher elevations. Similarly, the small mammal fauna was distinct from that of the lowlands with more and different species of non-volant mammals on the mountaintop. Three families of water beetles new for Suriname were recorded on the Grensgebergte. Two snakes not recorded from the lowlands were also recorded on the mountaintop.

In addition, the two lowland sites, the Juuru Camp (Upper Palumeu River, Site 1) and Kasikasima (Site 4) both had high levels of biodiversity but showed some differences in species composition. Only 50% of the water beetle fauna was similar between the two sites. Many species of small mammals (especially bats), reptiles and amphibians, and birds were recorded from only one of these two lowland sites. Fish species composition differed between these sites due to habitat differences- smaller creeks and thus smaller fishes were documented in the Upper Palumeu while larger fishes were collected in the main channel of the Palumeu River at Kasikasima and in Makrutu Creek (Site 3). A 50 m waterfall in the Upper Palumeu region contained several fish species new to science and to Suriname.

TAXONOMIC SUMMARIES

Water Quality

Fourteen sites on the Palumeu River, Tapaje Creek and Makrutu Creek revealed typical water quality conditions for undisturbed aquatic ecosystems in Suriname's interior, except for mercury. Although analysis of rainwater samples taken on the Grensgebergte did not have any mercury, levels above international norms were occasionally found in sediment and fish tissue samples. This indicates that there might be an external mercury source. Further monitoring is needed to confirm this. This aspect is very important and needs immediate action because these headwaters provide drinking water and food for many local communities downstream.

Plants

Out of 609 plant specimens collected, 354 species, 152 genera and 93 families have so far been determined. At site 1, along the upper Palumeu River, we collected 188 plant specimens. At the Grensgebergte (site 2) we collected 69 plant specimens and 75 at the Makrutu camp (site 3). We also collected 11 plant specimens at the Palumeu village and 27 specimens at the rapids of the Palumeu River. We found 15 new plant species records for Suriname and two new genera. Two of these belong to lianas, four to shrubs and herbs and ten to trees. The surroundings of the Grensgebergte and the Kasikasima Mountains contain several vegetation types that are dominant and floristically distinct for the central and southern parts of Suriname. These vegetation types include tall dryland tropical forest on laterite/granite hills, short savannah (moss) forest and open rock vegetation, including rocky outcrops around rapids, and tall seasonally flooded forest. Within these vegetation types, we recorded nearly all of the 15 new plant species records and the two new genus records for Suriname. We also recorded several rare species with only a few known occurrences in Suriname and/or in the Guianas. The noteworthy species include several rare orchids that are listed in Appendices I & II of CITES, some carnivorous plants, and three tree species that are listed on the IUCN Red List, including one tree species listed as Critically Endangered. Plot surveys (0.1 ha) also indicated that the forests of South Suriname are floristically distinct from those of North Suriname, but do not significantly differ in tree alpha diversity. The forests on the Guiana Shield basement complex are not uniform as stated by some. Our findings indicate the pristine status of the forests and vegetation types in Southeastern Suriname, and the fact that these forests are still poorly explored.

Water Beetles

More than 2500 specimens of water beetles were collected representing 157 species in 70 genera. Twenty-six species and 8 genera are confirmed as new to science, with an additional 10–15 species likely to be undescribed. Surprisingly, more species were recorded here than during the Kwamalasamutu Region RAP in Southwestern Suriname despite less collecting effort. Additionally, there was a high species turnover between these RAP sites: 40% of the species recorded here were not found in the Kwamalasamutu Region. The families Lutrochidae, Hydroscaphidae, and Torridincolidae are recorded from Suriname for the first time. While a broad range of habitats contributed to the high species and lineage diversity, hygropetric habitats on granite outcrops in particular provided a wealth of new and interesting taxa.

Dung Beetles

Dung beetles are among the most cost-effective of all animal taxa for assessing biodiversity patterns, yet RAP's recent surveys are among the few that are expanding our knowledge of Suriname's little known dung beetle fauna. In addition to cost-effective sampling using standardized pitfall traps, dung beetles depend upon large mammals for food and consequently can be used to rapidly assess the health of the overall mammal community and hunting impacts in a fraction of the time it would take to survey the mammals themselves. I sampled dung beetles using baited pitfall traps and flight intercept traps in the Grensgebergte and Kasikasima regions of Southeastern Suriname. I collected 4,483 individuals represented by 107 species. This ranks among the most diverse places on the planet for dung beetles, and exceeds the extraordinarily high species richness observed in nearby southwestern Suriname (94 species, Larsen 2011). Ten species are most likely new to science, while an additional 10–20 species may be undescribed pending further taxonomic revisions.

Dung beetle species richness, abundance and biomass were higher around Upper Palumeu than at Kasikasima, probably due to the extensive intact forest and lack of hunting pressure in this remote headwater region where no one currently lives. Dung beetle diversity and abundance at Kasikasima were still relatively high, indicating only mild to moderate hunting of large mammals and birds in the region. All sites, including the Grensgebergte Mountains, supported high endemism, including several rare species, demonstrating the exceptional biodiversity value of the region. Surprisingly, dung beetle species composition varied strongly among sites within this survey, as well as among sites sampled on previous surveys, including nearby southwestern Suriname. This high Beta diversity shows that the forests of Suriname and the Guiana Shield are not nearly as homogenous as is often assumed, and consequently protecting this varied biodiversity requires conserving many different forest areas.

The high abundance of several large-bodied dung beetle species in the region is indicative of the intact wilderness that remains. These species support healthy ecosystems through seed dispersal, parasite regulation and other processes. Maintaining continuous primary forest and regulating hunting (such as through hunting-restricted reserves) in the region will be essential for conserving dung beetle communities and the ecological processes they sustain. These results indicate that the intact headwater region of the Upper Palumeu watershed merits the highest conservation priority.

Katydids

Fifty-two species of katydids were collected, representing 35 genera in 4 subfamilies. At least 6 species are new to science and 26 species were recorded for the first time from Suriname; one of the new species recorded during the survey represents the first example in the Neotropical region of the loss of the ability to produce sound in male katydids. The katydid fauna of this country exhibits a remarkably high turnover—50% of species recorded during the current survey had not been collected during the 2010 survey in the Kwamalasamutu region, and all represent records new to Suriname. This may indicate a potentially high degree of uniqueness of the Grensgebergte mountains and warrants their protection.

Ants

A total of 149 ant species from 35 genera have been identified from the RAP collections. Additional work is ongoing to process and identify the remaining samples, which will undoubtedly raise the total number of species, possibly to over 200 species. The results indicate a healthy and diverse ant fauna reflective of pristine rainforest. Ants play important roles as predators, scavengers, and seed-dispersers in tropical forests. The ant data from Southeastern Suriname will add to a growing dataset on the ant fauna of the Guiana Shield, which is still poorly documented, to help identify areas of high diversity and endemism that are important to conserve within the region. Data on ants and other invertebrates are important since these groups may be able to illustrate differences between habitats within the Guiana Shield that larger animals with wide geographical ranges do not discern.

Fishes

A total of 94 species of fishes were recorded which, in combination with a collection of fishes from Lower Palumeu River by Covain et al. (2008), makes a total of 128 species now known to occur in Palumeu River. This diversity is high compared to the rest of the world, but is typical for a medium-sized river of the Guiana Shield. Eleven species of fishes are potentially new to science, including a *Bryconops* species, a small *Parotocinclus* catfish, and a tetra (*Hemigrammus* aff. *ocellifer*). Two species are new records for Suriname: *Hyphessobrycon heterorhabdus* and *Laimosemion geayi*; a third and fourth species, *Ituglanis nebulosus* and *Pimelodella megalops*, may also represent new species for Suriname if their identity is confirmed. We did not find the same species at each site; sites 3 and 4 included large-sized fishes from the main channel of the Middle Palumeu River, while site 1 had many small-sized species of creek habitat. Overall, large top level predators were still common in Palumeu River.

Reptiles and Amphibians

A total of 47 species of amphibians and 42 species of reptiles were recorded during the RAP survey. These numbers are relatively high when compared with other sites sampled over the same time period (e.g., recent RAP surveys in other parts of Suriname). Seven (six frogs and one snake) of the total 89 species encountered could not be assigned to any nominal species. These unidentified taxa may represent novel species yet require validating genetic and morphological data before formal diagnoses can be made. A number of records represent range expansions for taxa within the Guiana Shield (e.g., *Rhinatrema bivitattum*, *Alopoglossus buckleyi*).

Additionally, a teiid lizard (*Cercosaura argulus*) was recorded for just the second time in Suriname. Encountering >80 total species (including 19 snake species) is evidence of a healthy, diverse and seemingly pristine forest ecosystem.

Birds

A total of 313 bird species were seen or heard at the three RAP survey sites, the village of Palumeu, and during excursions along the Palumeu River. We recorded fourteen species listed as Vulnerable or Near-Threatened on the IUCN Red List, and consider another seven species as likely to occur in the region. Our records of several species represent range extensions within Suriname and the Guiana Shield. Whereas the lowland forest avifauna was broadly similar at the different localities, 32% of species were only observed at one of the four survey sites. The abundance of parrots and cracids was particularly noteworthy, especially compared to the more populated Kwamalasamutu region that we surveyed in Southwestern Suriname in 2010. The high-elevation savanna forest harbored several species not known to occur in the adjacent lowlands, and therefore had the most unique species assemblage of any site. Our results indicate that the lowland forest of SE Suriname probably contains the vast majority of bird species known to occur in the country's interior, including many species of high conservation value, arguing strongly for protection of the region's forests. We recommend further surveys of high-elevation sites in the Grensgebergte and other mountain ranges in southern Suriname, to better determine the range limits of species restricted to high-elevation forests.

Small Mammals

A total of 39 species of small mammals (<1 kg) were documented during the RAP survey. Taxonomic composition included 28 species of bats, 8 species of rats, and 3 species of opossums. Of the 3 sites sampled, the lowland sites were most similar, with Upper Palumeu having the highest diversity of bats and Kasikasima having the highest abundance for bats. The highland site of Grensgebergte had the highest diversity and abundance for small non-volant mammals but the lowest for bats. The species composition was heterogeneous with no opossums shared among sites, whereas 25% of rats and just over 50% of bats were shared among sites. The most noteworthy records were the documentation of the poorly known water rat (*Nectomys rattus*) near the open granite outcrop of Grensgebergte. This region of Southeastern Suriname has a mix of primary rainforest in a mosaic of lowland and highland habitats that supports diverse and different faunal communities of small mammals.

Large Ground Dwelling Mammals

During the survey of large and medium-sized ground dwelling mammals of the Kasikasima and Upper Palumeu river region, we recorded 18 species. Camera traps were the most important tools for the survey, but direct observations were made and tracks, scat and scratch marks were also recorded. We observed important species, including endangered and vulnerable species, such as Jaguar, Tapir and Giant River Otter. All these species fulfil important roles in the ecosystem such as by dispersing seeds or regulating populations of other species. The occurrence of a high diversity of large and medium-sized mammals in the surveyed area indicates that the ecosystem is healthy and relatively pristine. Southeastern Suriname is very important for large mammal species, especially wide-ranging species, because the area encompasses vast tracts of pristine forest and rivers.

Primates

Six of the eight primate species known from Suriname were recorded during the RAP survey. These included the black spider monkey (*Ateles paniscus*), red howler monkey (*Alouatta maconnelli*), bearded saki (*Chiropotes sagulatus*), brown capuchin (*Cebus paella*), squirrel monkey (*Saimiri scuireus*), and golden handed tamarin (*Saguinus midas*). The large-bodied species (black spider monkey, red howler monkey) were fairly abundant, indicating sustainable hunting practices by local communities. Although we did not record the white faced saki or wedge capped capuchin, they may occur in the area. These species are quite difficult to observe due to rarity and elusiveness. Primates play keystone ecological roles in maintaining healthy ecosystems, and also are important for local people. The high diversity and abundance of primates in the area make it a high conservation priority.

CONSERVATION RECOMMENDATIONS

(see also Chapter 1 for a discussion of the conservation importance of Southeastern Suriname)

We strongly recommend protection of Southeastern Suriname to preserve its unique and diverse species, as well as its forest and freshwater resources. Protection of Southeastern Suriname will:

- Protect unique biodiversity that is found nowhere else,
- Conserve critical natural resources for the well-being of Suriname and the world,
- Guarantee perpetuity of Suriname's freshwater resources—a prerequisite for economic activities (agriculture, energy, mining, oil) and human health (consumption, sanitation, transportation)—by protecting and conserving the headwaters of one third of Suriname's Rivers,
- Mitigate global climate change through the conservation of large tracts of carbon rich tropical forest,
- Ensure sustainable flow of forest resources (e.g., food, medicines, building materials) and freshwater for the indigenous and Maroon communities in the interior of Suriname, and
- Maintain large-scale ecological processes and protect wide-ranging species and species vulnerable to climate change through establishment of a vast network of international conservation areas.

Protect the freshwater resources of Southeastern Suriname
Water quality and flows
Water quality of Southeastern Suriname is currently very high and typical for undisturbed aquatic ecosystems in Suriname's interior. This region contains the headwaters of several of Suriname's largest rivers, including the Marowijne, supplying much of the water used downstream for drinking, agriculture, energy, mining, sanitation, transportation, etc. Thus, maintaining high quality freshwater for the people and natural ecosystems of Suriname is essential for long-term sustainability. While the area may not be directly affected by mercury pollution from downstream sites, the RAP results show that it does receive atmospheric mercury deposition probably from trans-boundary sources. Protecting the head-waters of Southeastern Suriname and minimizing mercury pollution from neighboring countries will be important for safeguarding this source of clean freshwater and protecting human health.

Our findings also show that Southeastern Suriname will be disproportionately important for future water resources for the country. Watersheds in Southeastern Suriname are predicted to be more resilient to climate change than other parts of Suriname, where water scarcity could become a problem (see Chapter 1 this volume).

Fishes
The fishes of Southeastern Suriname are a major food source for the local indigenous and Maroon communities. The primary threat to the fishes of the region is the Tapajai Project, which proposes to build one or more dams in the Tapanahony River in order to divert its water via Jai Creek to Brokopondo Reservoir and thus increase power generation by the hydroelectric station at Afobaka. The dam(s) would not only directly affect migratory fishes, fishes of running water and creek habitats and fishes downstream of the dam(s), but also effectively mix the fish faunas of the Marowijne River System and the Suriname River System, each with its own endemic species, likely leading to species extinctions. Local communities along the Tapanahony River should be extensively informed about the potential impacts of the Tapajai Project on their immediate environment so they can make rational, well-informed choices about their future with or without the Tapajai Project.

Conserving the forested headwaters and middle reaches of the Palumeu River will be important for maintaining food for people for many years to come. Large stretches of seasonally flooded forest and swamps predominantly occur in southern Suriname. These habitat types, as well as mountain headwaters, appear to provide important spawning grounds for a variety of fish species, including large migratory species which are one of the most important food sources for people throughout Suriname. The fishes of Palumeu River can also be of interest to the aquarium hobby and sport fishers and thus generate income for local people if catches are regulated.

Actions that should be taken to protect the fisheries:
1. Assess which fish species from the Palumeu River are used for food, determine the amount caught and eaten, and study their life histories to determine how fast they reproduce and grow
2. Determine the amount of fish that can be sustainably harvested, both for food fishes and aquarium fishes
3. Conduct more research to understand migratory fish behavior in Southeastern Suriname
4. Set catch limits and/or seasons if necessary to avoid overfishing
5. Create picture guides of fishes, especially colorful species and fun-to-catch fish species
6. Promote sustainable catch and export of aquarium fishes to generate local income and conservation incentives

Protect the unique habitats and biodiversity of Southeastern Suriname
Flora
Vegetation plot studies revealed that species composition of the forests in the South of Suriname is different from the North, and a small set of significant indicator species and genera were found for the South of Suriname. The RAP results indicate that forests on the Guiana Shield basement complex are not one uniform forest type as has been suggested. Some forest types like the tall dryland forests on laterite/granite hills are more dominant in the South of Suriname, and are floristically distinct.

The flora of southern Suriname is surprisingly varied, and contains at least sixteen distinct habitat types (see Chapter 1 this volume). The surroundings of the Grensgebergte and the Kasikasima Mountains contain several vegetation types that are dominant within southern Suriname and floristically distinct for this region. Within these vegetation types, we recorded nearly all of the fifteen new plant species and two new genera for Suriname we found during this study. Nine plant records including one genus new for Suriname were found in the hilly landscape and at higher elevations. The other six new plant records including one plant genus new for Suriname were found in seasonally flooded forest and swamp forest along the Palumeu River. We also recorded several species with a restricted distribution in Suriname and/or in the Guianas, orchids listed on Appendices I & II of CITES, some carnivorous plants and three tree species that are listed on the IUCN Red List. Amongst these is a unique palm species, *Syagrus stratincola*, that is only know from ten localities in the Guianas, and is listed as Vulnerable on the IUCN Red List. We also found the tree species *Vouacapoua americana* that is listed as Critically Endangered on the IUCN Red List. As plant collections from several plant families still await identification, we expect to find more new records or noteworthy species for Suriname. These findings indicate the pristine status of the forests and vegetation types in Southeastern Suriname, and the fact that these forests are still poorly explored.

Fauna

As with plants, the differences in species composition for most animal taxa between this expedition and the 2010 Kwamalasamutu RAP survey, combined with the high turnover between camps within Southeastern Suriname, suggest very high Beta diversity and species turnover within southern Suriname. Many species appear to be very localized or patchily distributed, which leads to high overall diversity over relatively small areas. Forests of southern Suriname and of the Guianas in general are often considered relatively homogenous, but our findings contradict this pattern. Because of this high heterogeneity, conserving a wide range of species requires conserving numerous large areas of forest that only superficially appear to be the same.

The findings of dozens of species new to science on both RAP surveys, while not unexpected given the paucity of collecting in the region, reinforces how much further we have to go before we have an understanding of the biological diversity of southern Suriname. The collection of eleven potentially new fish species in the Upper and Middle Palumeu River under unfavorable (high-water) conditions during the present study and the absence of many fish species from the rapids in the present collection both indicate a richer fish fauna in Palumeu River than the fish fauna that is currently known (128 species). In order to arrive at a more complete list of the fish fauna of Palumeu River the following actions are recommended:

- Additional scientific surveys are necessary to document the fish biodiversity at different times of the year, but especially when river levels are lower; and collection efforts should be aimed mainly at the major rapid complexes and the main river channel and tributaries in middle reaches of the Palumeu River
- The 1966-collections of King Leopold III of Belgium and J.P Gosse from the Palumeu River (Table 8.2) should be studied

Hunting

The Grensgebergte area is quite pristine, and is visited only occasionally by the local people while they travel from the villages in southern Suriname to villages in Brazil. Hunting does not seem to pose a threat for the large and medium-sized mammals in the Grensgebergte area, because the area is so remote and very inaccessible. We found no signs of hunting or any other human disturbance. The large bodied primates (black spider monkey and red howler monkey) are present in relatively high abundance. They were either spotted or heard on a regular basis at both sites. Since these two species are the most hunted by local communities, this indicates sustainable hunting practices by these communities. Southeastern Suriname provides a refuge for many species that are heavily hunted in other parts of the Guiana Shield, as well as a source of food as these animals reproduce and disperse into hunted areas.

In the Kasikasima area there are more human activities than in the Grensgebergte area, such as ecotourism and some hunting by a small number (10–20) of people that live in Kampu. Kampu is a small settlement along the Middle Palumeu River a few kilometers downstream of the Papadron rapids. Currently the greatest potential threat for large and medium-sized mammals at Kasikasima would be hunting, but the results from the RAP survey show that hunting pressure is currently low. Nevertheless, the presence of species sensitive to hunting and disturbance such as Jaguar, Puma, Tapir, curassows and large primates also suggests that hunting pressure is low, and that the important ecological processes maintained by these species, such as seed dispersal and population regulation, remain intact. Dung beetles, which are often used as a rapid indicator of hunting pressure, also suggest low levels of hunting, since they were diverse and abundant during the survey. However, dung beetle species richness and abundance were still lower around Kasikasima than around Grensgebergte, reflecting these differences in hunting pressures.

Hunting is probably limited by reduced river access to some areas in the dry season, and more generally by distance from Palumeu and other villages. The absence of a market and the concentration of the indigenous people in Palumeu both reduce hunting pressure on large vertebrates in the region as a whole. The extensive surrounding forest acts as a source to offset local population depletion due to hunting. Nevertheless, the most significant current threat to medium- and large-bodied mammals in the area is hunting from Palumeu village. This could change if plans to build a road from northern to southern Suriname move ahead. A road would make the area accessible and hunting and habitat destruction would become important threats for large terrestrial vertebrates. Further plans to increase hydro energy capacity of the existing hydro lake in Brokopondo by diverting water from the Tapanahony River to the Jai Creek could be a threat if the project would take place, because a part of the area in South Suriname will be flooded. Aquatic mammals, such as giant otters, may be particularly vulnerable. More data on large mammal populations will be necessary to manage these possible future threats. Recommended studies include more camera trapping and a sustainability evaluation of wild bushmeat hunting to have better baseline data.

The value of extensive wilderness and ecosystem services

Southeastern Suriname is very important for wide-ranging species, such as large mammals, birds and fish because the area encompasses vast tracts of pristine forest and rivers. In fact, there are few places left on earth that are as pristine as Southeastern Suriname. Many large mammal and bird species have broad home ranges and can move freely across Southeastern Suriname in the absence of disturbance. Large migratory fish species move long distances to spawn. The area is also connected to protected areas in Brazil to the south and to a national park in French Guiana to the east. This makes Southeastern Suriname part of a large wilderness area, which is important to sustain genetic diversity within large mammal species, as well as other taxa such as reptiles and amphibians. The RAP sites are most likely acting as a corridor for gene flow through this region of the Guiana Shield. The presence of species that are rarely seen or were previously unrecorded in Suriname helps to substantiate that there is (or was) an historical connection between this and surrounding areas. Maintaining the pristineness of this corridor should be a priority for healthy ecosystem function and to maintain natural gene flow throughout the Guiana Shield.

This contiguous network of protected areas, with Southeastern Suriname at its center, also allows species to persist in the face of climate change by providing corridors for redistribution. This is especially important as many species are being forced to move upslope from the lowlands into the mountains as climate warms. In addition to helping species adapt to climate change, the extensive, intact and carbon-rich forests of Southeastern Suriname also help to regulate regional climate and mitigate global climate change.

Furthermore the area is important for the production of ecosystem services directly used by humans, such as water, food, medicines, recreation and lands of indigenous people (see Chapter 1 this volume). Since a large part of the rivers in Suriname originate in this area, protecting Southeastern Suriname guarantees flows of fresh water which is used for economic activities downstream such as transportation, hydro energy, agriculture and mining. The area also has a high potential for ecotourism, because of the beautiful pristine landscape and rich biodiversity, particularly of charismatic birds and mammals.

As one of the last remaining wilderness areas and a key provider of ecosystem services, we believe it is essential to protect Southeastern Suriname and the numerous benefits the region provides to people throughout Suriname and the world, avoiding threats from large-scale projects such as roads, mining and hydropower in this part of the country.

REFERENCES

Alonso, L.E., J.L. Deichmann, S.A. McKenna, P. Naskreki and S.J. Richards (editors). 2011. Still Counting… Biodiversity Exploration for Conservation—The First 20 Years of the Rapid Assessment Program. Conservation International, Arlington, VA, USA, 316 pp.

Huber, O., and M.N. Foster. 2003. Conservation priorities for the Guayana Shield: 2002 Consensus. Conservation International, Washington, D.C.

O'Shea, B.J., L.E. Alonso, and T.H. Larsen (eds.). 2011. A Rapid Biological Assessment of the Kwamalasamutu region, Southwestern Suriname. RAP Bulletin of Biological Assessment 63. Conservation International, Arlington VA.

Short, A.E.Z. and V. Kadosoe. 2011. Aquatic beetles of the Kwamalasamutu region, Suriname (Insecta: Coleoptera). *In:* O'Shea, B.J., L.E. Alonso and T.H. Larsen (eds.) 2011. A Rapid Biological Assessment of the Kwamalasamutu region, Southwestern Suriname. RAP Bulletin of Biological Assessment 63. Conservation International, Arlington, VA.

Chapter 1

Importance of Conserving Southeastern Suriname

Armand Moredjo, Lisa Famolare, Leeanne E. Alonso and Trond H. Larsen

As the greenest country on Earth, Suriname has long protected its forest resources through sound conservation management. Suriname's status as a high forest low deforestation (HFLD) country, its immense freshwater resources, its high biodiversity, rich tropical ecosystems, and low population density place the country in a truly unique position to become a global model for sustainable development and to take advantage of emerging ecosystem service markets and natural capital valuation schemes.

Natural and Cultural Resources of Suriname

- Suriname is located within a global treasure: the largest tract of pristine rainforest on Earth

- Suriname has extensive freshwater resources that are essential for the country and are a future resource for the region and the world

- Many Indigenous and Maroon communities in Suriname rely on the forest and freshwater resources for their livelihood and survival

---

Suriname is located within the Guiana Shield, a vast tropical wilderness covering over 2.2 million square kilometers in northern South America and containing over 25% of the world's tropical rainforests in the largest tract of pristine rainforest on Earth. With most of its 539,000 human inhabitants residing along the coast, Suriname maintains 95% of its original forest cover, comprising 148000 km² of pristine lowland rainforest, savanna, and montane ecosystems. Nowhere else on Earth exist such extensive, unaltered forests filled with an incredible diversity of animals and plants, inhabited by few human beings. Suriname and its forests contain high biodiversity with over 740 species of birds (Ribot 2013), 207 species of mammals (IUCN 2013), 104 species of amphibians (Ouboter 2012) and and 481 currently known fresh- and brackish-water fish species (Mol et al. 2012). Suriname is also blessed with plentiful supplies of high quality ground and surface freshwater, ranking globally among the top 10 nations of the world in renewable freshwater resources (FAO Aquasat Data, The World Bank 2012). Seven major watersheds capture and carry freshwater throughout the country (Table 1.1).

Table 1.1. The 7 major watersheds of Suriname.

Surface area of watershed (km²)		Average discharge (m³/sec)	
Corantijn	67,600	1,597	Joint estuary: 1,771
Nickerie	10,100	174	
Coppename	21,700	565	Joint estuary: 882
Saramacca	9,400	257	
Suriname	16,500	422	Joint estuary: 591
Commewijne	6,600	169	
Marowijne	68,700	1,791	

In addition to natural resources, Suriname is also rich in cultural diversity. A great diversity of people live along the coast, with origins from India, the Netherlands, Indonesia, China, several African countries, and many other nationalities. The interior of Suriname is inhabited principally by several indigenous groups, including the Wayana and Trio, and Maroon tribes, descendants of escaped African slaves who live mostly in the country's interior.

Suriname is one of the last places on Earth where an opportunity still exists to conserve extensive tracts of pristine diverse tropical forests and freshwater. Suriname has an annual deforestation rate of 0.02%, exhibits the lowest population density of any moist tropical region on Earth (0.2 people/ha), has few roads in the forested part of the country (which can be accessed only by small boat, small plane, or on foot), has 29.6 hectares of forest per capita and virtually all of the lands are public and under the control of the national government or indigenous and Maroon communities.

However, the isolation that has protected Suriname's ecosystems, natural resources, and indigenous cultures is disappearing at an increasing rate, and the opportunity to act to preserve these remarkable resources will soon be gone. Record high commodity prices have encouraged the rapid growth of small-scale gold mining activities as illustrated by a 192% increase in gold production (kg) from 2000 to 2011. Timber production increased by 107% from 2000 to 2011, with an additional 10–20% growth expected over 2012. These changes are also providing incentives to press ahead with new infrastructure projects.

As global access to freshwater and forest resources decline, Suriname's resources will only become more valuable. Conserving these remarkable resources now will provide Suriname a unique opportunity to develop sustainably, to mitigate climate change, and ensure the country's sustained economic growth and prosperity.

IMPORTANCE OF SOUTHEASTERN SURINAME

Southeast Suriname is the most isolated and pristine region of the country, and perhaps the world. Most of the region is uninhabited by humans, with only a few small villages of indigenous and tribal peoples to the north and east. The area is mostly covered by lowland rainforest, partially inundated along the many rivers, and on terra firme (unflooded) ground at higher elevations, which ranges from 25–900 m above sea level. Scattered throughout the region are many granitic rock outcrops (inselbergs) that rise above the forest canopy. The region also contains several mountain ranges including the Grensgebergte in the southwest, the Tumukhumak Mountains in the southeast, and the Oranjegebergte in the center (See Map 1, page 13).

Southeastern Suriname is critical to the health and wellbeing of Suriname

- Protects a major source of freshwater for the country, the region, and the world (used for food, transport, energy, agriculture, mining, etc.)

- Provides a sustainable supply of forest resources for the people of Suriname (e.g. water, food, medicines, and recreation)

- Ensures long-term resilience of Suriname's freshwater resources despite predicted freshwater declines elsewhere in response to climate change

- Maintains potential for future economic and water-based infrastructure development in Suriname

- Mitigates global climate change through conservation of large tracts of carbon rich tropical forest

- Safeguards an exceptional diversity of species and healthy ecosystems

- Represents high potential for sustainable economic growth through ecotourism—one of the world's largest growing industries

---

Ecological and Biological Importance

Southeastern Suriname supports exceptionally rich biodiversity, making it a high priority region for conservation. This region was highlighted in the Guiana Shield Priority Setting Workshop held by CI, IUCN and UNDP in April 2002 (Huber, Foster 2002) as one of the highest ranked areas for biodiversity conservation (Conservation International 2011).

The area is contiguous with the Tumucumaque Indigenous Reserve (3,071,070 million ha), the Tumucumaque National Park (3,877,393 ha) in Brazil and the Parque Amazonien de Guyane (2 million ha) in French Guiana (see Map 6, page 17). Consequently, ensuring the protection of this area is crucial for maintaining connectivity between this much larger network of protected areas spanning three separate countries. This type of large-scale ecological connectivity is essential for maintaining broad ecosystem services such as regional climate regulation, and healthy and genetically diverse populations of wide-ranging species, such as jaguar and migratory fish. The contiguous network of protected areas also allows species to persist in the face of climate change by providing corridors for redistribution, especially from the lowlands into the mountains.

South Suriname includes at least 16 land cover types identified through remote sensing of Landsat satellite imagery combined with field observations (Bánki and Aguirre 2011, Map 3, page 15). These land cover types include, for example, flooded forest, mixed dryland rainforest, granite forest, shrub forest, savannahs and wetlands (see Bánki and Aguirre 2011 for more details).

The 2012 RAP survey revealed that the area of the Upper Palumeu River around the Grensgebergte and Kasikasima mountains contains a wealth of biological diversity. Due to its extensive forest, remote location, and pristine nature, the area contains species not commonly seen in other parts of Suriname or elsewhere in the Neotropics, such as large cracid birds (guans, curassows) and macaws, jaguar, puma, and eight species of primates. These species are more heavily hunted in other parts of the Guiana Shield, but are thriving in the undisturbed refuge of Southeastern Suriname.

The RAP results also show that the Grensgebergte Mountains harbor species not found elsewhere in Suriname and may contain unique species assemblages of birds, small mammals, reptiles, ants and beetles. The plant communities of the region differ from those of northern Suriname (see Chapter 3 this volume), the water beetle and dung beetle fauna differs from other sites sampled in southwestern Suriname (see Chapters 4 and 5 this volume), and the bird and small mammal communities of the Grensgebergte differ from the surrounding lowlands (see Chapter 10 and 11 this volume). Over 50 species new to science were documented in Southeastern Suriname during the 2012 RAP survey (see Executive Summary this volume) and many more remain to be discovered. In addition to supporting unique species, Southeastern Suriname contains very high overall species richness relative to other regions, which is likely influenced by the diversity of habitat types, including elevational gradients in the mountains, and by the lack of historical disturbance.

Freshwater Importance

The rivers originating in the Grensgebergte, Eilerts de Haan gebergte, Tumukhumak, and Oranjegebergte feed into the upper Marowijne River and the Tapanahony River which joins other rivers to form the 68,700 km² Marowijne River watershed. This watershed provides freshwater for transportation, food, drinking and bathing for ca. 15,000 people in the region including French Guiana as well as to ca. 35,000 people downstream along the Marowijne river and along the eastern coast and as far as Paramaribo. The western portion of the area also likely feeds water into the Suriname River and Brokopondo Lake which generates hydro power for the country.

The freshwater resources of Southeastern Suriname are a critical source of water for the local communities living along the rivers. Protection of the Southeastern Suriname watersheds would ensure flows of fresh and clean water far into the future. Impacts of small scale mining on water quality are already apparent downstream, with serious consequences for human health. Consequently, avoiding upstream mining is especially important to reduce health impacts, as mercury may be deposited widely via atmospheric deposition (see Chapter 2 this volume). While the 2012 RAP survey concluded that the Palumeu River and its tributaries have high water quality conditions typical of undisturbed aquatic ecosystems in the interior of Suriname, mercury analysis of sediments and fish tissues indicates that some mercury contamination is entering the area (see Chapter 2 this volume). Protecting the headwaters of Southeastern Suriname will be important for safeguarding this source of clean freshwater for local communities and the entire country.

The fish of the rivers of Southeastern Suriname are a critical source of protein to the indigenous and Maroon people living along the Tapanahony, the Palumeu and Suriname rivers. Fish are a common and highly valued food source. Large and medium-sized fish species that are routinely eaten have local names. Large fishes like anyumara (*Hoplias aimara*) and kwimata (*Prochilodus* and *Semaprochilodus*) are popular food fishes and are also most vulnerable to overfishing. While not eaten, many of the small fish species are highly valued in the aquarium hobby and could play a beneficial economic role in the development of the area if fisheries for these species are strictly regulated (see page 23).

Although the 128 fish species documented during this RAP and a previous survey in the Palumeu River is comparable to the number of species collected in other rivers of the Guiana Shield, the species composition is distinct from the Coppename River and the Sipaliwini River (see Chapter 8 this volume). Seven fish species collected during the RAP survey are endemic to the Marowijne River system, with one of these species, *Aequidens paloemeuensis*, is endemic to the Palumeu River proper (see Chapter 8 this volume).

A variety of economically and locally important fish species are migratory, and spawn near the headwaters or in the upstream flooded forests of Suriname's rivers. Since the mountains of Southeastern Suriname support many of these headwaters and flooded forest habitats, it is likely that they play in important role in sustaining spawning grounds for migratory fishes. Several species encountered during the 2012 RAP survey, such as *Prochilodus*, *Semaprochilodus*, and *Pseudoplatystoma tigrinum*, are likely to migrate there to spawn. Consequently, conserving the forests and rivers of Southeastern Suriname is essential to ensure food security for Suriname's tribal and indigenous people locally, as well as to protect the migratory species that people throughout Suriname depend upon.

Southeastern Suriname, Freshwater and Climate change

As is the case in many countries around the world, climate change threatens the long-term sustainable economic development of Suriname and is likely to affect many poor and indigenous communities disproportionately. Using an assemblage of all climate models to explore the value of Southeastern Suriname in terms of climate resilience, Southeastern Suriname emerges as one of the most climate resilient places in the country and therefore, its protection is essential to Suriname's sustainable development and successful adaptation to climate change (See Map 5, page 16). The mountainous regions of Southeastern Suriname are of particularly high value for ensuring sustainable flows of water as climate is changing. Effectively managing the watersheds in Southeastern Suriname will ensure greater long-term climate resilience by conserving forests, biodiversity, freshwater, agriculture and water infrastructure for future generations and will be one of the boldest climate adaptation strategies for Suriname, which is both low cost and high reward.

Cultural Importance

Research conducted by CI-Suriname indicates that the local indigenous communities in Southeastern Suriname depend heavily on the forest and freshwater resources of the region to sustain their livelihoods (Map 4, page 15). The forests, rivers, creeks, mountain ranges and savannahs are critically important for their economic well-being, culture, recreational pursuits, subsistence (hunting, fishing, growing cassava, gathering wood and nuts, etc.) and necessary for their future generations and their continued way of life. Particularly important for most villages are the rivers and other freshwater resources, which act as key modes of transport, provide critical food resources and support physical and mental wellbeing through their cultural significance and role in hygiene and sanitation (Map 4, page 15).

REFERENCES

Bánki, O.S. and J. Aguirre. 2011. Mapping the unexplored forests of Suriname. Report of a pilot study to develop a land cover / vegetation map for Southern Suriname. Internal report of the University of Amsterdam and Conservation International.

Conservation International. 2011. Review of the Guiana Shield Priority Setting Outcomes: Narrative Report. Report produced for the United Nations Development Programme, Georgetown. 79p.

FAO Aquasat Data, The World Bank. Accessed 2012. Renewable Internal Freshwater Resources per Cubic Meter. http:// data.worldbank.org/indicator/ ER.H2O.INTR.PC/countries/1W?display=default

Huber, O. and M.N. Foster. 2003. Conservation Priorities for the Guiana Shield: 2002 Consensus. Conservation International, Washington, DC, USA.

Mol, J.H.A. 2012. The Freshwater Fishes of Suriname. Volume 2 of Fauna of Suriname. Brill Academic Pub. The Netherlands.

Mulligan, M. *Submitted*. "WaterWorld: a self-parameterising physically-based model for application in data-poor but problem-rich environments globally." Submitted to Hydrology Research, 2012.

Ouboter, P.E. and R. Jairam. 2012. Amphibians of Suriname. Volume 1 of Fauna of Suriname. Brill Academic Pub. The Netherlands.

Ribot, J.H. 2013. Checklist of the birds in Suriname, South America. Reviewed by Jan Hein Ribot, May 2011, with the help of Otte Ottema and Arie L. Spaans. Updated July 2013, by Jan Hein Ribot. http://www. surinamebirds.nl/php/listbirds.php. *Accessed July 12, 2013*.

Sáenz, L. 2012a. WaterWorld. A web-based modeling system to explore the impact of land use, climate change and land management interventions on hydrological services. Factsheet. Moore Center for Science and Oceans, Conservation International, Arlington, VA, USA.

Sáenz, L. 2012b. Nature and climate resilience of Suriname. Working paper. Conservation International, Arlington, VA, USA.

Chapter 2

Water quality assessment of the Palumeu River

Gwendolyn Landburg

SUMMARY

During this assessment we sampled water quality at 14 sites on the Palumeu River, Tapaje Creek and Makrutu Creek. In general the result show typical water quality conditions for undisturbed aquatic ecosystems in Suriname's interior, except for mercury. Although analysis of rain water samples taken on the Grensgebergte did not have any mercury, levels above international norms were occasionally found in sediment and fish tissue samples. This indicates that there might be an external mercury source. Further monitoring is needed to confirm this. This aspect is very important and needs immediately action because these headwaters provide drinking water and protein source for many local communities downstream.

INTRODUCTION

Although most of Suriname's rivers have their origin in the south of Suriname, little is known about water quality near the headwaters of these south-north meandering rivers. Most of these source rivers drain the extensive mountain range that forms the south border of Suriname. One of these rivers, assessed during this RAP survey, is the Palumeu River, draining the mountainous area in Southeastern Suriname. The main river springs from the Grensgebergte mountains, while big and small tributaries of the Palumeu River drain other parts of the Grensgebergte (for example Tapaje Creek), Toemoek hoemak gebergte (for example Makrutu Creek) and Oranje gebergte (Makrutu creek). Baseline water quality data for this area is very limited. Assessment of water quality in this area is not only important for ecological reasons, but also to identify if the water is safe to be used as drinking water for villagers within and downstream from the area assessed.

Parameters to measure were selected mainly to obtain a general impression of water quality in the area, and to test for the effects of disturbance. Anthropogenic disturbance was not expected, because of the position and remoteness of the assessed area, (Ouboter et al. 2012). Ouboter et al. (2012) predicted that due to the NE trade wind mercury is transported SW inland, causing mercury deposition on the lee side of the mountainous area in mid West Suriname. According to this prediction both the NW and the SE areas would have minimum mercury deposition. Parameters (for example turbidity and metals) were selected to confirm this.

SITES

Sampling was conducted around three major sites on the Palumeu river (in total 14 sites; Figure 2.1): upstream Palumeu river near the Grensgebergte (7 sites), one site on the Grensgebergte, upstream to midstream Palumeu river (4 sites), and midstream Palumeu river (2 sites).

The upstream portion of the Palumeu River is narrow and creek-like, with a maximum width of 20 m, shallow clear water, white sand, strong meandering, overhanging vegetation, swamp vegetation, and seasonally flooded creek forest. Upstream to midstream, the Palumeu River gets wider and less meandering with steep walls and high dry land forest. Tapaje Creek is structurally similar to the midstream portion of the Palumeu River, with steep walls and high dry land forest, while Makrutu Creek has a lot of moko moko swamp forest downstream along the creek. Downstream of the Tapaje and Makrutu Creek the Palumeu River starts to have rapids and rocky forest islands.

METHODS

We measured 11 physico-chemical parameters at each site: pH, dissolved oxygen, conductivity, temperature, alkalinity, total phosphate, nitrate, chloride, tannin & lignin, ammonia and turbidity (Appendix 2.1). At selected sites, water samples were saved for later analysis for mercury, iron, and aluminum at the University of Suriname in Paramaribo. Sediment and fish tissue samples were taken opportunistically for mercury analyses. All stored samples were kept under refrigeration in the field. Titrimetric, spectrofotometric and colorimetric methods were used to assess the parameters.

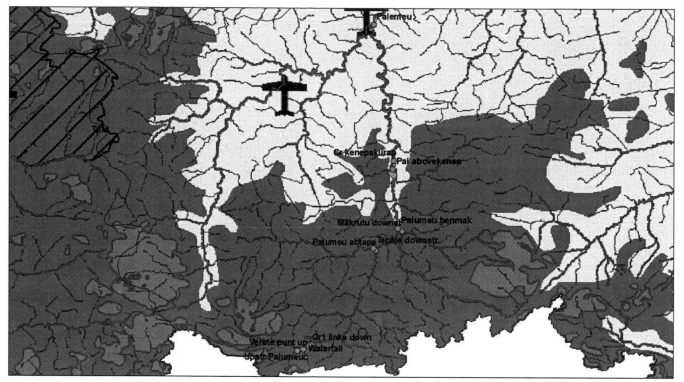

Figure 2.1. Water quality sampling sites during the 2012 RAP survey.

RESULTS

The results of the assessed parameters can be found in Appendix 2.1. In general, the creeks of the Palumeu River show lower oxygen levels (1.72 – 7.93 mg/L) than the main river (5.58 – 7.34 mg/L) mainly due to the lack of rapids in the creeks. Due to organic input from the forest, nutrient concentrations are higher in the creeks (phosphate: max 0.13 mg/L; ammonia: 0.09 mg/L; nitrate: 0.02 mg/L) than in the river (phosphate: max 0.11 mg/L, ammonia: 0.08 mg/L, nitrate: 0.01 mg/L). Both creeks and river have relatively clear water (max turbidity: 11 NTU, which was measured after rain). Concentrations of both aluminum and iron were found in creeks (aluminum: 0.33 mg/L average, iron: 0.80 mg/L average) and rivers (aluminum: 0.36 mg/L, iron: 0.59 mg/L). The results of the analysis of rainwater, gathered on top of the Grensgebergte granite outcropping, showed that there is deposition of aluminum and iron with rain in the area. Mercury concentrations in water were low in both creeks (0.00 – 0.06 μg/L) and river (0.02 – 0.06 μg/L), and there was no mercury found in the Grensgebergte rainwater. Mercury levels in sediment were higher in creeks (0.16 μg/g average) than in the river (0.19 μg/g average). For fish, mercury levels measured in *Hoplias aimara*, (N=4) caught in the Palumeu River (furthest point upstream) were on average 0.59 μg/g and in *Serrasalmus rhombeus* (N=1), caught in the Makrutu Creek, 0.36 μg/g.

DISCUSSION AND CONCLUSIONS

Compared to other sites, the data imply typical water quality conditions for undisturbed aquatic ecosystems in the interior for all creeks and river sites assessed (O'Shea et al. 2011, Alonso and Berrenstein 2006, Ouboter 1993), except in the case of mercury. No mercury was detected in rainwater, at least at the level of precision of hundredths of micrograms per liter, which implies that deposition of mercury with rain in the area is minimal. compared to other pristine sites, for example the Tafelberg in the Central Suriname Nature Reserve, where 0.11 μg/L was found in rainwater (unpublished data). This is in line with the hypothesis proposed by Ouboter et al. (2012) that mercury is being transported inlands by the northeast trade wind and deposited by rain in the mid-West southern mountain ranges.

Consequently, high levels of mercury occur in rivers in southwestern Suriname even in remote, undisturbed areas such as around Kwamalasamutu (Landburg and Hardjoprajitno 2011). Consistent with this hypothesis, we did not expect to find mercury pollution in the current study sites which are located farther to the east, out of the Norteast trade wind direction. Mercury levels around Kwamalasamutu were in some cases more than two times higher than those found here. The watersheds of Southeastern Suriname appear to be sheltered from this phenomenon and may support one of the cleanest freshwater in all of the country. Nonetheless, compared to the global background levels for mercury in sediment of 0.01 – 0.05 μg/g (Anderson

1979), the mercury levels found in the samples are still high. Mercury levels in fish were on average slightly higher than the European Union standard for human consumption of 0.5µg/g (EC 2002) for the Palumeu River but lower for the Makrutu Creek. Although these upper watershed sites are isolated from any potential impacts of mining further downstream, they do appear to be affected by atmospheric mercury deposition; probably from transboundary sources. This area may be out of position to be affected by mercury pollution from downstream sites but is not spared from transboundary mercury pollution.

Protecting the headwaters of Southeastern Suriname and minimizing mercury pollution from neighboring countries will be important for safeguarding this source of clean freshwater for especially the traditional communities more downstream who are using this water as drinking water and protein source.

REFERENCES

Alonso, L.E.. and H.J. Berrenstein (eds). 2006. A Rapid Biological Assessment of the Aquatic Ecosystems of the Coppename River Basin, Suriname. RAP Bulletin of Biological Assessment 39. Conservation International, Washington DC.

Anderson, A., 1979. Mercury in soils. *In:* J.O. Nriagu (ed.). The biochemistry of tertiary volcanic rocks in parts of the Virginia City Quadrangle. Amsterdam, Elsevier. Pp. 70–112.

EC. 2002. EC Regulation (221/2002) amending Commission Regulation (EC) no. 466/2001 of 8 March 2001 setting maximum levels for certain contaminants in foodstuffs. EC, Brussels.

Landburg G. and Hardjoprajitno M. *In:* O'Shea B.J., Alonso L.E. and T.H. Larsen (eds). 2011. A Rapid Biological Assessment of the Kwamalasumutu region, Southwestern Suriname. RAP Bulletin of Biological Assessment 63. Conservation International, Arlington VA.

O'Shea, B.J., L.E. Alonso and T.H. Larsen (eds). 2011. A Rapid Biological Assessment of the Kwamalasamutu region, Southwestern Suriname. RAP Bulletin of Biological Assessment 63. Conservation International, Arlington VA.

Ouboter, P.E, G.A Landburg, J.H.M. Quik, J.H.A Mol, and F. van der Lugt. 2012. Mercury Levels in Pristine and Gold Mining Impacted Aquatic Ecosystems of Suriname, South America. Ambio. 41(8): 873–882.

Ouboter P. (ed), 1993. The Freshwater Ecosystems of Suriname, first edition, Kluwer Academic Publishers, 313 p.

Appendix 2.1. Water quality of the Palumeu River.

Parameter	1	2	3	4	5	6
Location name	Palumeu river, upstream Palumeu Camp	Waterfall creek across Palumeu camp	Creek 1 downstream	Creek 1 upstream	Creek 2 upstream camp	Furthest point upstream Palumeu river
Coordinates	21 N 652292	21 N 652731	21 N 653902	21 N 651429	21 N 650745	21 N 650466
	273868	271528	274917	273890	273444	273281
Location description	River, 15–20 m width, trunks in the river, vegetation over the water	Creek, ± 60–70 m high; clear water, rocks; sandy substrate in creek	Creek, clear water, sandy substrate, slow current, trunks in the creek, vegetation over the creek	Creek, ± 25 m width, vegetation over the creek, steep walls, trunks in the water, fast current	Creek, vegetation over the creek, clear water, slow current, trunks in the water, steep walls	River gets the character of a creek, 5-10m width, sandy bottom, clear water, fast running, shallow, vegetation over the creek
Date	10/3	11/3	12/3	13/3	14/3	14/3
Time	17:00	11:00	10:59	11:30	12:30	14:45
Weather condition	cloudy	sunny	cloudy, sunny	sunny	cloudy, sunny	cloudy, sunny
pH	5.5	5.9	5.6	5.7	5.6	6.0
Conductivity (µS/cm)	20	20	15	14.1	17.2	18.2
Temperature (°C)	23.9	23.5	22.9	22.6	22.7	22.9
Dissolved oxygen (mg/L)			1.72	7.22	7.42	7.34
Dissolved oxygen (%)			20	83.4	86	85.2
Alkalinity (mg/L CaCO3)	5.2	4.75	3.5	5.3	5.5	6.65
Tanin Lignin (mg/L)	5.1	2.6	0.9	5	3.6	3.325
Phosphate (mg/L)	0.05	0.03	0	0.07	0.13	0.11
Nitrate (mg/L)	0.01	0.01	0.01	0.02	0.02	0
Ammonia (mg/L)	0	0	0.04	0.07	0.07	0.08
Turbidity (NTU)	7	5	4	11	7	9
Chloride (mg/L)	5.7	3.2	3.6	5.5	3.5	3.4
Aluminium (mg/L)	0.32	0.28	0.27	0.38	0.33	0.31
Iron (mg/L)	0.82	0.33	0.52	1.1	0.79	0.83
Mercury water (mg/L)	0.03	0.03	0.03	0	0	0.02
Mercury sediment (µg/g)	0.16		0.15	0.17	0.16	0.16
Mercury fish average (µg/g)						0.59
Remarks	rain during the day					

table continued on next page

Appendix 2.1. *continued from previous page*

Location #	Location name	Coordinates	Location description	Date	Time	Weather condition	pH	Conductivity (µS/cm)	Temperature (°C)	Dissolved oxygen (mg/L)	Dissolved oxygen (%)	Alkalinity (mg/L CaCO3)	Tanin Lignin (mg/L)	Phosphate (mg/L)	Nitrate (mg/L)	Ammonia (mg/L)	Turbidity (NTU)	Chloride (mg/L)	Aluminium (mg/L)	Iron (mg/L)	Mercury water (mg/L)	Mercury sediment (µg/g)	Mercury fish average (µg/g)	Remarks
7	Creek 2 downstream camp	21 N 654466 275245	Creek, 10–20 m width, very slow current, abundant vegetation in and over the creek, deep; clear water	15/3	10:30	very cloudy	5.6	19.3	23.1	4.58		6.5	0.2	0.04	0	0.09	9	4.5	0.32	2.31	0.06	0.17		rain in the night
8	Tapaje creek downstream	21 N 673367 303638	Creek, ± 100 m width; clear water, fast current; terra firme forest	19/3	10:45	cloudy sunny	5.3	11.7	23.6	6.40		1.9	6.7	0	0.02	<0.008	6	3	0.38	0.39	0.06	0.27		
9	Downstream Palumeu above Tapaje	21 N 674285 302969	River, fast current, above rapids, clear water	19/3	12.01	cloudy sunny	5.6	13.7	23.7	5.58	66	2.2	2.5	0.07	0.01	<0.008	11	3.6	0.35	0.41	0.06	0.15		
10	Makrutu creek downstream	21 N 682107 308705	Creek, 60–100 m width, floodplains, moko moko vegetation, clear water, fast current	20/3	10:45	cloudy	5.6	15.3	23.8	5.31	63	2.8	4.3	0.03	0	0.07	8	4	0.29	0.54	0.06		0.36	
11	Palumueu river downstream Makrutu creek	21 N 680949 309517	River, above rapids, fast current, rocks in the water	20/3	10:57	cloudy	5.5	13.1	23.7	5.64	67	2.2	2.4	0.09	0	0.03	9	3.5	0.45	0.43	0.02			rain in the night
12	Creek Kenepaku rapid, downstream camp	21 N 679570 329530	Creek, 5–10 m, rocky substrate, vegetation in and over the water, very slow current	22/3	9.05	cloudy sunny	5.9	15.7	23.2	6.59		3.95	0.9	0.06	0	0.01	2	1	0.35	0.37	0.02			
13	Palumeu river upstream of Kenepaku camp	21 N 679320 326922	River, fast current, clear water	23/3	9:30	cloudy sunny	5.5		23.9	7.93	94	2.05	2.2	0.06	0	0.03	6	3	0.37	0.45	0.03			
14	Rainwater Grensgebergte rock		Granite outcropping	14/3	9:00	cloudy, rain													0.31	0.05	0			

Chapter 3

Floristic assessment of the Upper Palumeu River, the Grensgebergte, and the Kasikasima areas

Olaf Bánki and Chequita Bhikhi

SUMMARY

We collected a total of 609 plant specimens during the RAP survey, including 433 fertile and 175 sterile plants. The majority, 238 plant specimens consisting of more than 50 percent of sterile collections, was collected in the surroundings of Kasikasima (site 4). Of these specimens, 602 were identified to family level, 512 to genus and 439 to species. The identified plant specimens belong to 354 species, 152 genera and 93 families. At site 1, along the upper Palumeu River, we collected 188 plant specimens. At the Grensgebergte (site 2) we collected 69 plant specimens and 75 at the Makrutu camp (site 3). We also collected 11 plant specimens at the Palumeu village and 27 specimens at the rapids of the Palumeu River. We found 15 new plant species records for Suriname and two new genera. Two of these belong to lianas, four to shrubs and herbs and ten to trees. The Grensgebergte and the Kasikasima Mountains contain several vegetation types which are dominant and floristically distinct for the central and southern parts of Suriname. These vegetation types include tall dryland tropical forest on laterite/granite hills, short savannah (moss) forest and open rock vegetation, including rocky outcrops around rapids, and tall seasonally flooded forest. Within these vegetation types, we recorded nearly all of the 15 new plant species records and the two new genus records for Suriname. We also recorded several rare species with only a few known occurrences in Suriname and/or in the Guianas. The noteworthy species include several rare orchids that are listed on appendices I and II of CITES, some carnivorous plants, and three tree species that are listed on the IUCN Red List, including one tree species listed as Critically Endangered. Plot surveys (0.1 ha) also indicated that the forests of South Suriname are floristically distinct from those of North Suriname, but do not significantly differ in tree alpha diversity. The forests on the Guiana Shield basement complex are not uniform as stated by some. Our findings indicate the pristine status of the forests and vegetation types in Southeastern part of Suriname, and the fact that these forests are still poorly explored.

INTRODUCTION

Plants are the principle building blocks of forests, and the fundamental components of ecosystems. On a global and national level, forests supply vital ecosystem services that sustain life on earth. Forests also support livelihood of millions of people throughout the developing world (Hall 2012). Forests in tropical regions, however, despite their importance are under pressure of global change (e.g. mostly human impacts). The countries in the Guianan region still contain large stretches of pristine tropical rainforest due to their low human populations. Together with the Amazonian forests, the Guianan forests belong to the largest area of pristine tropical rainforest in the world. Suriname, especially the southern region of the country, is the least botanically explored compared to the other countries in the Guianan region. Not much is known in terms of plant composition and diversity from the South of Suriname.

During the 2010 Rapid Assessment Program (RAP) survey around Kwamalasamutu in south-western Suriname, we recorded 9 plant species as new records for Suriname, including a newly recorded plant species for the Guianan region, and many rare and endangered species (Bánki and Bhikhi 2011). The knowledge that can be gained through this rapid assessment is needed to assess the conservation value of the forests in southern Suriname. This conservation value is needed for sound decision-making and conservation planning in the southern Suriname. During the 2012 three-week RAP survey reported here, we studied the flora of the Grensgebergte and the Kasikasima region of Southeastern Suriname. This report presents the preliminary results of the plants that were collected and inventoried.

METHODS

The floristic team consisted of approximately 10 members: André Semmy (tree spotter, plot inventories), Jeffrey Krimbo (field assistant plant collecting, plot inventories), about seven field assistants and the authors of this chapter for general plant collecting, plot inventories, and species identification.

We sampled at all four RAP sites: Upper Palumeu River (Juuru Camp, Site 1), Grensgebergte Granite Rock (Site 2), Makrutu Creek (Site 3), and Kasikasima camp (Site 4). At each of these sites we carried out general vegetation surveys. At Site 1 and Site 4 we carried out plot inventories. We collected for a period of nineteen days.

General Vegetation Surveys

General plant collecting took place at each of the four RAP sites in all the vegetation types encountered. This included collecting plants on the rocky outcrops of the Grensgebergte and the Kasikasima, along river edges and on rocky outcrops along rapids between the RAP sites, at the Palumeu Mets Lodge, and in Palumeu village. Flowering and fruiting plants were collected during the surveys, and all encountered vegetation types were recorded. Plant collections were collected in the number series of Olaf Bánki and Chequita Bhikhi (using OSB and CRB in plant collecting numbers). Plants were pressed and dried in the field above kerosene stoves. All plants were collected where possible in three duplicates, with one duplicate stored at the National Herbarium of Suriname (BBS), one stored at the National Herbarium of the Netherlands - Naturalis Biodiversity Center (NHN-NBC) and where relevant one duplicate send to a plant specialist for identifications. Leaf materials of plots were collected on silica to enable identification by DNA fingerprinting techniques, allowing phylogenetic classification of the plots in the future. Some plants and plant parts were collected on alcohol. Plants were identified by Chequita Bhikhi and Olaf Bánki at the NHN in Leiden, where plant collections of the Guianas are stored. The specimens were identified using standard identification techniques. Duplicates of 5 plant families were sent to their respective taxonomic group specialist, of the Herbier de Guyane (CAY) in Cayenne French Guiana (Rubiaceae), Kew Royal Botanic Gardens (Kew, Chrysobalanaceae and Myrtaceae), Smithsonian National Museum of Natural History (Sapindaceae) and The New York Botanical Garden (NYBG, Burseraceae) within the Flora of the Guianas network. We determined new records for Suriname by checking the occurrence of the species in the checklist of the Guianas (Funk et al. 2007), the Guianas collections of the NHN, and by consulting the collections of the Missouri Botanical Gardens through Tropicos and Discoverlife (www.discoverlife.org), and the available species occurrence data from the Global Biodiversity Information Facility (www.gbif.org). We also made use of the Encyclopedia of Life (eol.org), and the plant list (www.theplantlist.org).

Plot Inventories

We inventoried in total four 0.1 ha plots in 'tall dryland tropical forest on laterite/granite hills' (see Table 3.1 for the plot metadata). The plots were positioned on forested laterite/granite hills of 350 to 400 meters above sea level. The forests in these plots were standing on relatively shallow soils, with the bed rock occurring at various depths. Although we classified the forests in the plots as 'tall tropical forest on laterite/granite hills' at times these forests showed floristic elements of 'savannah forests', especially at those places where the bedrock was closer to the surface. All 0.1 ha plots had a dimension of 10 by 100 meters, and all trees above 2.5 cm dbh were pre-identified by tree spotter André Semmy of Conservation International Suriname and Olaf Bánki in the field. Palms were included in the assessment, while lianas were not assessed in these plots due to time constraints. Each new tree species encountered in the plots was collected. Collections of the tree species were processed in a similar way as the plant collections of the general surveys. We assessed the amount of timber species within the plots by using the Surinamese forest law of 1992 (see www.sbbsur.org).

At the first RAP Site at the Upper Palumeu River, we inventoried a plot (Gre1) on a granite hill in 'tall tropical forest on laterite/granite hills'. In the same area we tried to inventory another plot, as well as in 'savannah forest' on the rock outcrop of the Grensgebergte (Site 2). These two plots could not be inventoried because of logistical constraints. Close to the Kasikasima camp (Site 4), we assessed three 0.1 ha plots (Kas2-4). These three plots were established in 'tall tropical forest on laterite/granite hills' at a geographical distance of several hundreds of meters from one another.

Plot comparisons

To investigate floristic and diversity differences, we compared the four 0.1 ha plots of the current study with 0.1 ha plot data from other parts of Suriname. For the comparisons, we used six 0.1-ha plot data from Olaf Bánki (6 plots from Gros Rosebel on the Guiana Shield basement complex in the Northern part of Suriname), and three 0.1-ha plots from Kwamalasamutu and surroundings (Bánki and Bhikhi 2011).

Plot Analyses

To enable analyses, the four 0.1-ha plot datasets were brought together with the plots from Gros Rosebel (Bánki unpublished) and the Kwamalasamutu surroundings (Bánki and Bhikhi 2011) into one dataset of 13 0.1 ha plots in

Table 3.1. Metadata for the plots established at each site during the RAP. N = number of individuals, S = number of species, Fa = Fisher's alpha.

Plot Name	Ha	Dimension	N	S	Fa	Latitude	Longitude
Upper Palumeu River Plot 1 (Gre1)	0.1	100 × 10 m	177	67	39.27	N2 28.904	W55 37.898
Kasikasima Plot 2 (Kas2)	0.1	100 × 10 m	209	76	42.96	N2 58.493	W55 23.429
Kasikasima Plot 3 (Kas3)	0.1	100 × 10 m	203	73	40.86	N2 58.507	W55 23.526
Kasikasima Plot 4 (Kas4)	0.1	100 × 10 m	191	63	32.81	N2 58.522	W55 23.197

total, consisting of 2369 individual trees, and 369 (morpho-) species in total. Differences in floristic composition between plots were investigated with the ordination technique Non-Metric Multi-Dimensional Scaling (NMS) with Relative Sörenson as floristic distance measure, 250 real and random-ized data runs, and 4-6 dimensions (NMS in PCORD 5, McCune and Grace 2002; McCune and Mefford 1999). We also performed a Detrended Correspondence Analysis on both the 0.1-ha plot dataset (DCA in PCORD 5, McCune and Grace 2002; McCune and Mefford 1999). We per-formed these ordinations for all species, for all species with five or more individual trees in total in the complete dataset, and by using genus data alone. In this report we show the NMS ordination results for the genus data set, and the spe-cies with 5 or more individual trees in the whole 13 0.1 ha plot data set.

We also ran species indicator analyses on the 0.1-ha plot dataset to investigate which species were responsible for the division of the plots in several floristic groups (in PCORD 5, Dufrene and Legendre 1997; McCune and Grace 2002; McCune and Mefford 1999). We did these indicator analy-ses for all species, for all species with five or more individual trees in total in the complete dataset, and by using genus data alone. In this report we show the ISA results for the genus data set, and the species with 5 or more individual trees in the whole 13 0.1 ha plot data set.

The tree alpha diversity of the plots was expressed as Fisher's alpha (Fisher et al. 1943). Fisher's alpha is a diversity index describing the relation between the number of indi-viduals and species in a plot. Differences in the averages of the number of species, number of individuals, and in Fisher's alpha were statistically tested through ANOVA.

RESULTS

General Observations and Notes on Plant Diversity
In the RAP survey area, we discerned at least nine groups of vegetation types (following Lindeman and Moolenaar 1959; Bánki and Bhikhi 2011):

1. Tall herbaceous swamp and swamp wood
These vegetation types are abundant along the Upper Palumeu River, along the Makrutu Creek, and sporadically occur along the Palumeu River. Herbaceous swamps are characterized by the sheer dominance of herbaceous species like *Heliconia* sp. or *Montrichardia arborescens*, and mostly occur in shallow black waters associated with a peat or mud layer. Swamp wood consists of solitary trees standing in the water, such as *Erythrina* sp., or can be a combination of a herbaceous swamp with shrubs, and solitary trees, such as *Triplaris surinamensis*, *Virola surinamensis*, *Tachigali panuren-sis*, and *Cecropia* sp. In meandering black water rivers espe-cially at the coastal areas in Suriname, extensive stretches of both herbaceous swamps and swamp wood can occur. Along the larger rivers the bands of herbaceous swamps and swamp

wood can be fairly narrow almost immediately transitioning into real tall swamp forests or seasonally flooded forests (see also Lindeman and Moolenaar, 1959).

Typical for the Upper Palumeu River and prominent in river bends is the tall herbaceous swamp vegetation domi-nated by 'palulu' (*Heliconia* sp.). In this herb layer *Costus arabicus*, *Calathea comosa* and cyper grasses such as *Scleria flagellum* and *S. microcarpa*, were found. The occurrence of *Montrichardia arborescens* is scarce in the Upper Palumeu River. Vines and lianas such as *Manettia coccinea*, *Cayaponia cruegeri*, *Mucuna urens*, *Cynanchum blandum*, *Dioclea virgata* and *Stigmaphyllon sinuatum* occurred frequently in the swamp vegetation. The shrub and tree layer of the herba-ceous swamp vegetation close to the camp of RAP site 1 con-sisted of *Bixa orellana*, *Senna alata*, *Inga nobilis*, *I. splendens*, *Conceveiba guianensis*, *Croton cuneatus* and *C. pullei*. Trees of *Senna alata*, *Inga nobilis*, *I. splendens* and *Guarea guido-nia* were dominant in the river bends and formed a closed canopy of up to 5 to 10 meters in height.

Solitary trees of *Triplaris surinamensis* occurred frequently in the swamp wood and herbaceous swamp vegetation along the Upper Palumeu River. *Virola surinamensis* trees were also recorded in the swamp wood. At some places the band of swamp wood was very narrow, almost immediately transi-tioning into seasonally flooded forests. Within these narrow bands of swamp wood we found *Triplaris surinamensis* and *Virola* trees accompanied by small clumps of *Bactris brong-niartii* palms, clumps of *Euterpe oleracea* palms, and *Cecropia* species. Tree species generally occurring in seasonally flooded forest such as *Alexa wachenheimii*, *Guarea guidonia*, *Dialium guianense*, *Genipa americana*, *Macrolobium bifolium*, and *Posoqueria longiflora* frequently occurred in the shrub and tree layers of these narrow bands of swamp wood, especially in the river bends.

Stretches of tall herbaceous swamp vegetation dominated by *Montrichardia arborescens* had a frequent occurrence in the Upper Palumeu River after the confluence with the Tapaje Creek. *Montrichardia arborescens* was also domi-nant along the Makrutu Creek. Several plant species were recorded in the shrub vegetation along the Makrutu Creek such as *Cissus erosa*, *Tabernaemontana sylvilitica*, *Annona hypoglauca*, *Strychnos guianensis*, *Combretum rotundifolium*, *Solanum pensile*, *Ipomoea squamosa*, *Vigna luteola*, *Gouania blanchetiana* and *Paullinia dasygonia*. In this vegetation we also found solitary trees of *Cecropia latiloba* and *Bactris brongniartii* growing in large clumps. *Croton pullei* trees were dominant along the Palumeu River and the Makrutu creek.

2. Tall Seasonally flooded forest
Seasonally flooded forest is a tall forest type that usually is standing directly or almost directly at the river edge. In composition the seasonally flooded forest does not differ substantially in species composition from tall tropical low-land rainforest on dryland (terra firme). At the river edges where the flooding is substantially more frequent tree species especially occur that are accustomed to some extensive

flooding. Where the river banks are higher, and the soils well drained tall tropical lowland rainforest on dryland (terra firme) occurs (see also Lindeman and Moolenaar, 1959).

Stretches of seasonally flooded forest were most extensive and frequent along the Upper Palumeu River. Most dominant trees along the river edge, where flooding is more frequent, were *Alexa wachenheimii, Elizabetha princeps, Guarea guidona, Zygia racemosa, Ceiba pentandra, Croton matourensis, Trichilia* sp., *Posoqueria longiflora* and species of Burseraceae and Annonaceae.

Along the Palumeu River at site 4 (Kasikasima) we found *Simaba orinocensis, Parinari campestris, Protium heptaphyllum subsp. heptaphyllum, Duguetia calycina, Licania laevigata, Homalium guianense, Elizabetha princeps*, and *Eschweilera pedicellata* as dominant species in the tree layer of the seasonally flooded forest. In the understory we found *Miconia minutiflora* and several species of Myrtaceae. The understory was covered with lianas and vines such as *Mikania micrantha, Passiflora costata, Paullinia capreolata, Bauhinia cupreonitens*, and *Coccoloba ascendens*.

Along the Palumeu River at site 4 (Kasikasima) we also found the newly recorded species *Paloue induta, Heteropterys orinocensis* and the noteworthy species *Taralea oppositifolia* in the tree layer of the seasonally flooded forest. The newly recorded species *Tovomitopsis membranacea, Prestonia cayennensis*, and *Platymiscium filipes* were found in the seasonally flooded forest along the Palumeu river.

3. Seasonally flooded palm swamp forest and creek forest

Patches of *Euterpe oleracea* palm swamp forests were dominant along the Upper Palumeu River upstream from our first camp. This forest type is dominant in the lowest parts of the coastal plain, characterized with an abundant growth of *Euterpe oleracea* (Lindeman and Molenaar 1959). From the helicopter in the Grensgebergte area, this forest type was frequently observed in the creek valleys between hills. Swamp forest consists of heavy clay soils which is inundated most of the year, and stays at least damp in the dry season (Lindeman and Molenaar 1959). Along the trail to Brazil (site 1) and the trail to the Kasikasima Mountain (site 4) we encountered patches of palm swamp forest in wet (or inundated) areas and close to creeks. These patches consisted of *Euterpe oleracea* trees mixed with *Geonoma baculifera, Heliconia acuminata*, and several Marantaceae species such as, *Hylaeanthe unilateralis, Ischnosiphon obliquus, Monotagma spicatum, M. secundum* and *Calathea elliptica* in the herbal layer. Dominant trees near creeks were *Spondias mombin, Tabebuia insignis, Macrolobium bifolium* and *Duroia aquatica*. Wet patches dominated by *Hylaeanthe unilateralis, Costus scaber, Ischnosiphon obliquus, Heliconia densiflora* and the noteworthy species *Costus lanceolatus subsp. pulchriflorus* were found at site 4.

Along the creek at site 1, close to the waterfall we recorded *Macrolobium bifolium* trees, and *Psychotria racemosa, Justicia secunda , Gurania bignoniacea, Hyptis lanceolata* and fern species such as *Salpichlaena volubilis* and *Selaginella parkeri* in the understory on the banks.

In a creek bed along the trail to the Mets camp at camp 4 we found a different set of species. The soil consisted of loam mixed with fine white sand and was inundated. The herb layer below a somewhat open canopy was covered with *Rapatea paludosa* (Rapateaceae), *Miconia bracteata, M. ceramicarpa, Calathea elliptica, Heliconia richardiana, Olyra obliquifolia*, and an ants housing shrub with red fruits *Maieta poeppigii*.

Ischnosiphon hirsutus is newly recorded herb species for Suriname, and was found together with *Hylaeanthe unilateralis* in a palm swamp forest close to the waterfall at site 1.

4. Tall tropical rainforest on dryland (terra firme)

This forest type was only found at the Upper Palumeu River (site 1) along the trail to Brazil. This forest type can be distinguished from other forest types by its high canopies reaching to 50 m, and the well-drained soil. We observed trees such as *Alexa wachenheimii, Parkia pendula* with purple flowers, *Couratari stellata, Ormosia* sp., *Tabernaemontana sananho, Tetragastris altisimma*, and *Lecythis zabucajo* reaching a canopy height of more than 30 m, and *Astrocaryum sciophilum* (bugru maka) palms dominant in the understory. We frequently encountered *Psychotria poeppigiana, P. apoda, Heliconia lourteigiae, Heliconia densiflora, Costus scaber, C. claviger, Duguetia calycina, Maieta guianensis*, different fern species, and *Monotagma secundum* and *M. spicatum* as the most dominant Marantaceae species in the understory. Unfortunately we could not properly assess this forest type with plot inventories due to logistics.

5. Tall dryland tropical forest on laterite/granite hills

'Tall dryland tropical forest on laterite/granite hills' can be shortly described as a terra firme forest type that occurs on the slopes and the tops of laterite/granite hills. This forest type is a mix between tree species occurring in 'tall tropical rainforest on dryland', and those tree species that respond to the more shallow soils caused by parent rock under the surface. At those places where on the slope or on the top of the hills the parent rock reaches close to the surface, this forest type can become a mix with 'savannah forest on granite rock'. On the slopes where the soils are deep, the canopy can be high and uniform. At those places on the slopes and on the top where the parent rock under the surface start to fluctuate leading to patches of shallow and deeper soils, the canopy becomes more erratic in height at short distances and becomes more open at times allowing more light to penetrate the forest floor.

Tall dryland tropical forest on laterite/granite hills was the most dominant forest type found around the Upper Palumeu River at site 1 and the trail to the Kasikasima Mountain. For this reason we placed all plots in this vegetation type. The whole area at these sites started to become hilly at approximately 10 meters from the river.

At both RAP sites 1 and 4 we recorded characteristic tree species such as *Astrocaryum sciophilum*, *Astrocaryum paramaca*, *Eperua falcata*, *Eschweilera corrugata*, *Geissospermum sericeum*, *Inga* spp., *Licania* spp., *Pouteria guianensis*, *Protium* spp., *Tetragastris* spp., *Unonopsis glaucopetala*, *Zygia racemosa* and *Vouacapoua americana*. Other dominant trees species found at the Upper Palumeu River site were *Siparuna decipiens* and *Sterculia* sp.. The understory of the forest contained small trees and shrubs such as *Connarus fasciculatus*, *Siparuna cuspidata*, *Retiniphyllum* sp., *Psychotria* sp. *Ischnosiphon obliquus*, *Rinorea amapensis*, *Henriettella caudata*, *Piper bartlingianum*, *Maieta guianensis*, and the following dominant understory palms *Geonoma leptospadix*, *G. maxima* and *Bactris acanthocarpa*.

At the Kasikasima site 4 the following tree species were found: *Brosimum guianense* (letterhout), *Dicorynia guianensis* (basralokus), *Elizabetha princeps*, *Eschweilera* sp., *Guatteria schomburgkii* (pedrekupisi), *Inga alba* (rode prokoni), *Jacaranda copaia* (gubaja), *Laetia procera*, *Martiodendron parviflorum* (witte pintolokus), *Minquartia guianensis* (alata udu), *Oenocarpus bacaba*, *Parkia nitida*, *Pouteria* sp., *Pseudopiptadenia suaoveolens* (pikinmisiki), *Sclerolobium melinonii* (djakidja), *Sloanea* spp. (rafunyannyan), *Swartzia* sp. (zwarte bugubugu), *Talisia* sp., *Virola* sp., *Vochysia tomentosa* and *Zanthoxylum rhoifolium* (pritjari).

The following newly recorded species for Suriname were found (mostly) in the plot inventories: *Hirtella duckei* (plot 3), *Micropholis splendens* (plot 3), *Ouratea cerebroidea* (plot 3), *Quiina indigofera* (plots 1, 2 and 3), *Tabernaemontana angulata* (plot 3), *Tetrameranthus guianensis* (plots 2, 3 and 4).

Tall dryland tropical forest on laterite/granite hills was also found on the Grensgebergte granite mountain at an elevation of approximately 800 meters asl, but unfortunately due to logistics it could not be properly assessed. At the edges of this forest on the Grensgebergte we found two new recorded tree species for Suriname, namely *Clusia flavida* (see page 26) and *Solanum semotum*. The noteworthy species *Ixora piresii* was collected in the understory of the tall dryland forest a few meters from the trails of the Kasikasima Mountain site.

6. Short savannah (moss) forest on granite rock
Short savannah forest on granite rock was found on the top of the granite mountain of the Grensgebergte (site 2), along the rapids of the Palumeu River, in the surroundings of Kasikasima camp (site 4) and on top of the Kasikasima Mountains. This forest type is characterized by a short uniform canopy of several meters in height, many small tree stems, and a plant species composition that responds to the shallow soils on top of the granite rock. On the Grensgebergte Mountain and the Kasikasima Mountains (site 4) we observed low savannah forest covered in moss, potentially formed and sustained by the moisture from low hanging clouds. Typical tree species belonged to *Inga grandiflora*, *Myrcia bracteata*, *Myrsine guianensis*, *Symplocos guianensis*, *Byrsonima* sp. and to the families of Burseraceae (*Protium* spp.), Rubiaceae and Myrtaceae. *Inga stipularis* and *Miconia prasina* which occur along rivers or in inundated soils were also found on the Grensgebergte mountain.

At Kasikasima this forest type was most dominant at the base of the mountain, where large boulders occurred at the surface forming caves, small streams and pools, on top of the mountains, and along the steep slopes. Along the steep slopes and on the top of the mountain, we recorded a dominant IUCN Red Listed palm species *Syagrus stratincola*. This palm species also occurred on granite rocks along rapids of the Palumeu River. It was also observed in the bathing area of camp 4. At Kasikasima we also found a rare Myrtaceae species, *Eugenia tetramera* with large velvet haired yellow fruits in this forest type.

Except for *Syagrus stratincola* in the savannah forest on the rocks along the rapids of the Palumeu River we also found trees of *Cochlospermum orinocense*, *Topobea parasitica*, *Ouratea leblondii*, *Erythroxylum kapplerianum* and several species of Myrtaceae.

7. Open rock (Inselberg) vegetation, including rocky outcrops around rapids
Patches of open rock, and solitary rock outcrops such as Inselbergs have a distinct flora. The vegetation ranges from patches with shrubs and solitary trees to patches of herbs and solitary herbs such as cacti, bromeliads, Agavaceae or orchids. All plants occurring in this vegetation type are adapted to grow on the open rock surface. At places where there is a film of water flowing on the rock surface or where water is almost stagnant plants accustomed to nutrient poor conditions may occur. The same vegetation type can occur on rocky outcrops around rapids along rivers.

We encountered open rock vegetation on the Grensgebergte (site 2) and on the top of the Kasikasima Mountains. We also encountered open rock vegetation characterized by a gesneriad herb *Chrysothemis rupestris* on a large rocky slope under the canopy of the forest at site 1. This species was also recorded on the open rocks near the large rapids of the Palumeu river. These rock outcrops near large rapids were also characterized by open rock vegetation. We found some similar species among the rocky outcrops of the Grensgebergte, the Kasikasima Mountains and the rock outcrops near the rapids, irrespective of differences in elevation.

The herb layer on the Grensgebergte rock was dominated by a Bromeliaceae species, cyper grasses, an Apocynaceae shrub with white flowers, and orchids such as the *Epidendrum nocturnum*, *Maxillaria discolor*, *Phragmipedium lindleyanum*, and *Cleistes rosea*. In the herb layer, *Chelonanthus purpurascens* and *Turnera glaziovii* were also found. The shrub layer was made up by *Clusia* spp., Myrtaceae, Rubiaceae and Melastomataceae. In the shrub layer, the noteworthy species *Cavendishia callista* (Ericaceae) was encountered. This species is typical for the rock outcrops in the Guayana Highland region.

We observed several dominant species similar to the Grensgebergte and Kasikasima Mountains, such as the terrestrial Gesneriaceae with red flowers, *Sinningia incarnata*, the dominant Bromeliaceae, *Pitcairnia geyskesii* and many Cyperaceae species. *Sinningia incarnata* was also recorded on the rock outcrops near the rapids of the Palumeu River. *Anthurium jenmanii,* an Araceae species, was also dominant on the bare rocks of the Kasikasima Mountains. Species such as *Costus spiralis*, an orchid species and *Clusia* sp. were similar among the Grensgebergte Mountain and the rock outcrops at the rapids of the Palumeu river.

On the slopes of the Grensgebergte (site 2), and on the rock outcrops near the rapids of the Palumeu River we also encountered wet patches with carnivorous herbs of Lentibulariaceae (e.g. *Utricularia* spp.), grasses, other herbs and *Portulaca sedifolia*. The occurrence of carnivorous plants indicates the low nutrient status that some of these microhabitats have for plants.

We found a terrestrial, perennial Gesneriaceae herb *Lembocarpus amoenus*, with a restricted distribution in Suriname and French Guiana. The same species was also conspicuous on the bare rocks near the waterfall of site 1. This herb has only been found on ferro-bauxite Mountains and could be associated to rocky outcrops, and seems to grow on wet thin organic detritus layers in shaded areas.

As rocky outcrops around rapids are also treated in this section, and there is no separate vegetation description of plants within and around rapids in the water, we provide the following observation here. In the shallow water areas before rapids, where the water stream is not too turbulent, we found monodominant patches of small trees of *Psidium cattleianum* with reddish green leaves. This species was also dominant in shallow areas in the river, which will inundate at high water levels. Large monodominant patches of this species were also recorded along the Palumeu River, between camp 4 (METS camp) and the Palumeu village.

At the waterfall at site 1 we found the herb species *Clidemia epiphytica*, a new record for Suriname.

8. Secondary vegetation.
Secondary vegetation occurs on (abandoned) cultivated fields. It was observed along the trail to the Mets Camp and close to the Kasikasima camp (site 4). The vegetation contained secondary shrubs such as *Trema micrantha, Miconia bracteata, Solanum paludosum, S. subinerme, Lasiacis ligulata, Psychotria* sp., and trees of *Cecropia* sp. We recorded a former farm field close to camp four, with cultivated cassava plants (*Manihot esculenta*) belonging to the Trio village of the area.

9. Bamboo forest
Bamboo forest (*Guadua* sp.) was only observed in a small patch along the trail to Brazil at site 1. From the helicopter we observed patches of bamboo forest throughout the lowlands of the Grensgebergte area.

Plant collection, new records, and noteworthy plant species
We collected a total number of 609 plants during the RAP, including 433 fertile and 175 sterile plants (see Table 3.2).

The majority, 238 plants, mostly sterile, was collected at the fourth camp site.

Newly recorded genera and species for Suriname
According to the National Herbarium of the Netherlands (NHN-Leiden), the checklist of the plants of the Guiana Shield (Funk et al. 2007) and several other internet data portals (e.g. GBIF, Discoverlife, EOL, Tropicos) and checklists of plants (e.g. the plant list) we found 15 newly recorded plant species for Suriname during this expedition.

We also collected two genera that were recorded for the first time in Suriname. One, *Tetrameranthus,* is a rare genus within the family of the Annonaceae. The common characteristics within the Annonaceae are leaves positioned flat in a plane along a branch, and flowers that are divided into three parts. Within *Tetrameranthus* the leaf arrangement is spiral and flowers are divided in four parts (Westra and Maas 2012). Tetrameranthus guianensis is a newly recorded tree species for Suriname. We have collected the species in plots 2, 3 and 4 at RAP site 4 in 'tall dryland tropical forest on laterite/granite hills'. *Tetrameranthus guianensis* is recently described based on collections from French Guiana and the state of Amapá in Brazil (Westra and Maas 2012).

Tovomitopsis is the other newly recorded genus for Suriname, collected during this expedition. *Tovomitopsis membranacea*, is a tree species known within the Guianas from the states of Amazonas and Bolivar in Venezuela, French Guyana, and Guyana. The species occurs as well in western Amazonia (e.g. Colombia, Ecuador, Panama, Peru) (Funk et al. 2007, EOL 2013, GBIF 2013, Tropicos 2013). This tree was collected in the 'seasonally flooded forest' along the Palumeu River.

Solanum semotum (Solanaceae) is a newly recorded species for Suriname. We made our specimen collections from a tree of approximately 5 meters in height, collected at the edge of 'tall dryland tropical forest on laterite/granite hills' on top of the Grensgebergte granite mountain at an elevation of approximately 800 meters asl. *Solanum semotum* was previously collected in Brazil, French Guiana, and Guyana (Funk et al. 2007, GBIF 2013).

The next new species for Suriname was *Hirtella duckei* (Chrysobalanaceae) (see page 26), which was a small understory tree found in a gully on the trail to the Kasikasima Mountains. The tree species was also found in plot 3 at RAP site 4 (Kasikasima) in 'tall dryland tropical forest on laterite/granite hills'. The species is covered with brown hairs, and the leaf base bears swollen ant cavities which house mutualistic ants. The ants receive shelter from the plant, in turn protect the plant from herbivores. *Hirtella duckei* is known from central and western Amazonia (Brazil), Colombia, Peru and from Guyana, and also from the state of Amazonas in Venezuela (Funk et al. 2007, Prance 2007, Discover Life 2013, GBIF 2013).

Clusia flavida is the fifth newly recorded species for Suriname. This tree was also collected at the edge of 'tall dryland tropical forest on laterite/granite hills' on top of the

Table 3.2. List of plants collected along the Palumeu River, the Grensgebergte, the Kasikasimagebergte and surroundings during the CI Rap survey of 2012. Numbers indicate number of specimens collected at each survey site (River: Along the Palumeu River between sites, Site 1: Upper Palumeu River (Juuru Camp), Site 2: Grensgebergte Granite Rock, Site 3: Makrutu Creek, Site 4: Kasikasima camp, Village: Palumeu village).

* new records for Suriname; **IUCN Redlist species, *** species on CITES Appendices

FAMILY	SPECIES	RIVER	SITE 1	SITE 2	SITE 3	SITE 4	VILLAGE
Acanthaceae	Justicia secunda		1				
	Mendoncia bivalvis	1	1				
	Mendoncia cf. pedunculata		1				
Annonaceae	Anaxagorea dolichocarpa		1				
	Annona hypoglauca			1			
	Duguetia					1	
	Duguetia calycina		1			4	
	Duguetia riparia		1				
	Duguetia surinamensis					1	
	Guatteria punctata		1				
	Tetrameranthus guianensis*					1	
	Unonopsis					2	
	Unonopsis stipitata		1				
Apocynaceae	Ambelania acida					1	
	Aspidosperma cruentum					1	
	Geissospermum sericeum					1	
	Gonolobus ligustrinus					1	1
	Mandevilla	1					
	Mandevilla subspicata						1
	Mandevilla surinamensis						1
	Mesechites trifida			1			
	Pacouria guianensis		1				
	Prestonia cayennensis*					1	
	spp.			1		2	
	Tabernaemontana angulata*					1	
	Tabernaemontana sananho		1				
	Tabernaemontana siphilitica			1			
	Tabernaemontana undulata		1			1	
	Tassadia					1	
Araceae	Anthurium jenmanii					1	
	Anthurium rubrinervium		1				
	Colocasia	1					
	Monstera spruceana		1				
	Philodendron		1				
	Philodendron megalophyllum		1				
	Philodendron rudgeanum		1				
	sp.						1
Arecaceae	Bactris						1
	Bactris acanthocarpa var. acanthocarpa		1				
	Bactris brongniartii			1			

FAMILY	SPECIES	RIVER	SITE 1	SITE 2	SITE 3	SITE 4	VILLAGE
	Desmoncus polyacanthos			1			
	Geonoma leptospadix		1				
	Geonoma maxima		1				
	*Syagrus stratincola***					1	
Asclepiadaceae	*Cynanchum blandum*		1				
Aspleniaceae	*Asplenium laetum*					1	
Asteraceae	*Chromolaena odorata*					1	
	Mikania micrantha			1		1	
	Mikania parviflora			1			
	Piptocarpha						1
	spp.				1	1	
Balanophoraceae	*Helosis cayennensis*		1				
Begoniaceae	*Begonia glabra*		1				
Bignoniaceae	*Arrabidaea inaequalis*				1		
	Callichlamys latifolia		1				
	Distictella magnoliifolia		1				
	Mussatia					1	
	Paragonia pyramidata		1	2			
	spp.	1		1			
Bixaceae	*Bixa orellana*		1				
Blechnaceae	*Salpichlaena volubilis*		1				
Boraginaceae	*Cordia laevifrons*					2	
	Cordia tomentosa				1		
	Tournefortia cuspidata		1				
Bromeliaceae	*Aechmea mertensii*		1				
	Guzmania lingulata					1	
	Pitcairnia geyskesii						1
Burmanniaceae	*Apteria aphylla*						1
Burseraceae	*Protium*			1		5	
	Protium heptaphyllum ssp. *heptaphyllum*			1		1	1
	Tetragastris					3	
Cabombaceae	*Cabomba*				1		
Caryocaraceae	*Caryocar microcarpum*			1			
	Caryocar nuciferum					1	
Celastraceae	*Hippocratea volubilis*			1		1	
	Maytenus pruinosa			2			
Chrysobalanaceae	*Exellodendron barbatum*					1	
	Hirtella cf. *racemosa*		1				
	*Hirtella duckei**					1	
	Licania					5	
	Licania alba		1				
	Licania albiflora					1	
	Licania laevigata					1	
	Licania leptostachya			1			
	Parinari campestris					1	

FAMILY	SPECIES	RIVER	SITE 1	SITE 2	SITE 3	SITE 4	VILLAGE
Clusiaceae	*Clusia* aff. *flavida**						1
	Clusia leprantha			1			
	Clusia panapanari						1
	Rheedia macrophylla		1			1	
	Symphonia globulifera					1	
	Tovomita		1			1	
	Tovomita calodictyos		1				
	Tovomitopsis membranacea					1	
	*Tovomitopsis membranacea**				1		
Combretaceae	*Combretum*			1			
	Combretum laxum			1			
	Combretum rotundifolium			2			
	Terminalia amazonia					1	
Commelinaceae	*Dichorisandra hexandra*		1		1		
Connaraceae	*Connarus coriaceus*				1		
	Connarus fasciculatus		1				
	sp.					1	
Convolvulaceae	*Evolvulus alsinoides*			1			
	Ipomoea alba		1				
	Ipomoea tiliacea			1			
Costaceae	*Costus arabicus*		1				
	Costus claviger		1				
	Costus lanceolatus subsp. *pulchriflorus*					1	
	Costus scaber		1				
	Costus spiralis var. *spiralis*						1
Cucurbitaceae	*Cayaponia cruegeri*		1				
	Gurania bignoniacea		1				
	Gurania subumbellata		1				
Cyperaceae	*Calyptrocarya bicolor*		1				
	Cyperus			1			
	Diplacrum						1
	Rhynchospora barbata						1
	Rhynchospora cephalotes			1		1	
	Rhynchospora comata				1		
	Scleria				1		
	Scleria cyperina					1	1
	Scleria flagellum-nigrorum		1				
	Scleria microcarpa		1				
	Scleria stipularis		1				
	spp.	1	1			1	
Dennstaedtiaceae	*Lindsaea* cf. *parkeri*		1				
	Pteridium aquilinum						1
Dichapetalaceae	*Tapura amazonica*					1	
	Tapura guianensis		2			2	

FAMILY	SPECIES	RIVER	SITE 1	SITE 2	SITE 3	SITE 4	VILLAGE
Dilleniaceae	*Davilla kunthii*			1			
Dioscoreaceae	*Dioscorea*		1				
	Dioscorea cf. *debilis*			1			
Dryopteridaceae	*Tectaria incisa*		1				
	Triplophyllum dicksonioides		1				
Elaeocarpaceae	*Sloanea*					3	
Ericaceae	*Cavendishia callista*						1
Erythroxylaceae	*Erythroxylum*					1	
	Erythroxylum kapplerianum				1		
	Erythroxylum macrophyllum					1	
Euphorbiaceae	*Chaetocarpus schombugkianus*					1	
	Conceveiba guianensis		1			1	
	Croton cuneatus		1				
	Croton guianensis					1	
	Croton pullei		1	1			
	Euphorbia thymifolia	1					
	Sapium argutum						1
	sp.					1	
Fabaceae	*Alexa wachenheimi*		1				
	Bauhinia cupreonitens					1	
	Bocoa viridiflora					2	
	Candolleodendron brachystachyum					1	
	Chamaecrista nictitans var. *disadena*					1	
	Cynometra marginata					1	
	Dialium guianense			1			
	Dioclea coriacea			1			
	Dioclea elliptica		2				
	Elizabetha paraensis					1	
	Elizabetha princeps		1				
	Eperua falcata					1	
	Inga					3	
	Inga acrocephala		1			1	
	Inga brachystachys		1				
	Inga cf. *acrocephala*					1	
	Inga disticha			1			
	Inga graciliflora		1				
	Inga grandiflora						1
	Inga longiflora					1	
	Inga nobilis		4				
	Inga rubiginosa					1	
	Inga splendens		1	1			
	Inga stipularis						1
	Inga vera subsp. *affinis*			1			
	Inga virgultosa					2	

FAMILY	SPECIES	RIVER	SITE 1	SITE 2	SITE 3	SITE 4	VILLAGE
	Machaerium quinata			1			
	Machaerium trifoliolatum			1			
	Macrolobium bifolium		1				
	Mimosa myriadenia		1				
	Mucuna urens		1				
	*Paloue induta**					1	
	Phaseolus lunatus	1					
	*Platymiscium filipes**		1				
	Senna alata		1				
	Senna bicapsularis		1				
	Senna quinquangulata						1
	Senna silvestris var. *silvestris*			1			
	spp.	1				3	
	Swartzia					1	
	Swartzia benthamiana					1	
	Swartzia cf. *remiger*		1			1	
	Swartzia oblanceolata					5	
	Swartzia panacoco var. *polyanthera*			1		1	
	Taralea oppositifolia					1	
	Vigna juruana			1			
	Zygia latifolia var. *lasiopus*			1			
	Zygia racemosa		1				
Gentianaceae	*Chelonanthus purpurascens*						1
	Voyria clavata					1	
	Voyria rosea					1	
Gesneriaceae	*Besleria flavo-virens*		1				
	Chrysothemis rupestris		1				
	Codonanthe crassifolia		1				
	Drymonia coccinea		1				
	Drymonia serrulata		1				
	Lembocarpus amoenus		1				
	Nautilocalyx pictus		1				
	Paradrymonia campostyla		1				
	Sinningia incarnata					1	1
Haemodoraceae	*Xiphidium caeruleum*				1	1	
Heliconiaceae	*Heliconia acuminata*		1				
	Heliconia densiflora		2				
	Heliconia lourteigiae		1				
	Heliconia richardiana					1	
Hymenophyllaceae	*Hymenophyllum decurrens*						1
	Trichomanes vittaria					1	
Icacinaceae	*Discophora guianensis*					1	
indet			3			3	1

FAMILY	SPECIES	RIVER	SITE 1	SITE 2	SITE 3	SITE 4	VILLAGE
Lamiaceae	*Hyptis lanceolata*		1				
	Vitex orinocensis var. *multiflora*		1				
Lauraceae	*Licaria canella*					2	
	Ocotea guianensis					1	
	spp.		1			9	
Lecythidaceae	spp.		1			13	
Lentibulariaceae	*Utricularia*				1		1
	Utricularia subulata				1		
Linaceae	*Hebepetalum schomburgkii*					1	
Loganiaceae	*Spigelia hamelioides*		1				
	Strychnos			1			
	Strychnos guianensis			1			
Loranthaceae	*Oryctanthus alveolatus*			1			
	Phthirusa pyrifolia						1
Malpighiaceae	*Byrsonima spicata*						1
	*Heteropterys orinocensis**					1	
	Hiraea fagifolia		2				
	Stigmaphyllon convolvulifolium		1				
	Stigmaphyllon sinuatum	1					
	Tetrapterys crispa		1				
Malvaceae	sp.					1	
Marantaceae	*Calathea comosa*		1				
	Calathea elliptica		1				
	Hylaeanthe unilateralis		1				
	*Ischnosiphon hirsutus**		1				
	Ischnosiphon obliquus		1			1	
	Monotagma plurispicatum		1				
	Monotagma secundum		1				
	Monotagma spicatum		1				
Melastomataceae	*Aciotis indecora*		1				
	*Clidemia attenuata**						1
	Clidemia capitellata			1			
	Clidemia dentata		1				
	*Clidemia epiphytica**		1				
	Clidemia hirta			1			
	Ernestia granvillei					1	
	Henriettea patrisiana					1	
	Henriettea stellaris			1			
	Henriettea succosa			1			
	Henriettella caudata		2				
	Leandra rufescens					1	
	Macrocentrum fasciculatum						1
	Maieta guianensis		1				

FAMILY	SPECIES	RIVER	SITE 1	SITE 2	SITE 3	SITE 4	VILLAGE
	Maieta poeppigii		1				
	Miconia bracteata					1	
	Miconia ceramicarpa					1	
	Miconia ciliata						1
	Miconia lateriflora		1				
	Miconia minutiflora			1			
	Miconia prasina		1	1			1
	Miconia racemosa						1
	Miconia sagotiana						1
	Miconia serrulata		1				
	spp.	1	2			2	
	Topobea parasitica				1		
Meliaceae	*Guarea costata*					1	
	Guarea guidonia		2				
	Guarea pubescens subsp. *pubescens*		1	1			
	Guarea scabra					1	
	Trichilia		1			1	
	Trichilia schomburgkii subsp. *schomburgkii*					2	
Memecylaceae	*Mouriri*					1	
	Mouriri grandiflora		1				
	Mouriri vernicosa			1			
Menispermaceae	*Abuta grandifolia*		2				
	Abuta obovata				1		
	Orthomene schomburgkii					1	
Monimiaceae	*Siparuna cuspidata*					1	
Moraceae	*Brosimum guianense*					1	
	Ficus amazonica		1		1		1
	sp.					1	
	Trymatococcus amazonicus					2	
Myristicaceae	*Iryanthera*					1	
	Virola					1	
Myrsinaceae	*Cybianthus prieurii*		1			1	
	Myrsine guianensis						1
	Stylogyne atra		1				
Myrtaceae	*Calyptranthes*					1	
	Eugenia					1	
	Eugenia tetramera					1	
	Myrcia					2	
	Myrcia splendens					1	
	Psidium cattleianum					1	
	spp.		2	4	1	10	8
Nymphaeaceae	*Nymphaea glandulifera*		1				

FAMILY	SPECIES	RIVER	SITE 1	SITE 2	SITE 3	SITE 4	VILLAGE
Ochnaceae	Ouratea cerebroidea					1	
	Ouratea leblondi				1	2	
Olacaceae	Heisteria cauliflora					1	
	sp.		1				
Onagraceae	Ludwigia affinis					1	
	Ludwigia latifolia		1				
Orchidaceae	Cleistes rosea***						1
	Dichaea picta***		1				
	Epidendrum densiflorum***		1				
	Epidendrum nocturnum***						1
	Maxillaria discolor***						1
	Phragmipedium lindleyanum***						1
	spp.		2		1		
Parkeriaceae	Ceratopteris deltoidea			1			
Passifloraceae	Passiflora costata			1		1	
Piperaceae	Peperomia serpens		1				
	Piper anonifolium var. anonifolium		1				
	Piper arboreum		1				
	Piper bartlingianum		2				
	Piper demeraranum					1	
	Piper hostmannianum			1			
	sp.		1				
Poaceae	Lasiacis ligulata					1	1
	Olyra obliquifolia				1		
	Pariana radiciflora					1	
	spp.		3			2	
Polygalaceae	Barnhartia floribunda					1	
	Polygala adenophora						1
	Securidaca paniculata		1				
Polygonaceae	Coccoloba ascendens					1	
	Coccoloba parimensis						1
Polypodiaceae	Microgramma lycopodioides		2				1
	Microgramma persicariifolia			1			
	Niphidium crassifolium		1				
	Pecluma plumula				1		
	Pleopeltis percussa				1		
Portulacaceae	Portulaca sedifolia				1		
Putranjivaceae	Drypetes variabilis		1				
Quiinaceae	Quiina indigofera					1	
	Quiina indigofera*					1	
Rapateaceae	Rapatea paludosa					1	
Rhamnaceae	Gouania blanchetiana			1			
	Gouania velutina		1				

FAMILY	SPECIES	RIVER	SITE 1	SITE 2	SITE 3	SITE 4	VILLAGE
Rubiaceae	Alibertia edulis						1
	Borreria			1			1
	Duroia aquatica					1	
	Duroia eriopila		1				
	Faramea multiflora		1				
	Genipa americana			1			
	Isertia coccinea			1			
	Ixora piresii					1	
	Manettia coccinea var. spraguei		1				
	Palicourea		2			1	
	Palicourea longiflora						1
	Palicourea quadrifolia						1
	Palicourea triphylla						1
	Posoqueria					1	
	Posoqueria latifolia		1				
	Posoqueria longiflora		2				
	Psychotria		4		1	1	
	Psychotria bracteocardia					1	
	Psychotria hoffmannseggiana var. hoffmannseggiana						2
	Psychotria poeppigiana		1				
	Psychotria racemosa		1				
	Retiniphyllum		1				
	Sabicea			1			
	Sabicea amazonensis						1
	Sabicea oblongifolia			1			
	Sipanea pratensis						1
	Sipanea pratensis var. dichotoma				1		
	Spermacoce verticillata			1		1	
	spp.		1			2	
Rutaceae	Ertela trifolia				1		
Salicaceae	Casearia pitumba						1
	Homalium guianense					1	
	Ryania pyrifera					2	
Santalaceae	Phoradendron piperoides					1	
Sapindaceae	Paullinia capreolata					1	
	Paullinia dasygonia			1			
	sp.					1	
	Talisia					1	
	Talisia mollis		1				
	Thinouia myriantha		1				
	Toulicia cf. pulvinata					1	
	Toulicia pulvinata					1	

A Rapid Biological Assessment of the Upper Palumeu River Watershed (Grensgebergte and Kasikasima) of Southeastern Suriname

71

FAMILY	SPECIES	RIVER	SITE 1	SITE 2	SITE 3	SITE 4	VILLAGE
Sapotaceae	Micropholis cf. Splendens*					1	
	Micropholis guyanenis					3	
	Micropholis guyanenis subsp. guyanensis					2	
	Pouteria				1	3	1
	Pouteria cladantha					1	
	Pouteria guianensis					2	
	spp.					4	
Schizaeaceae	Anemia villosa						1
Selaginellaceae	Selaginella parkeri		1				
	Selaginella producta					1	
Simaroubaceae	Simaba cedron					1	
	Simaba orinocensis			1		1	
Siparunaceae	Siparuna cuspidata		2				
	Siparuna decipiens					1	
	Siparuna guianensis			1			
Solanaceae	Brunfelsia guianensis					1	
	Cyphomandra oblongifolia		1				
	Lycianthes pauciflora		1				
	Markea coccinea		1				
	Solanum crinitum	1					
	Solanum paludosum					1	
	Solanum pensile			1			
	Solanum rugosum	1					
	Solanum semotum*						1
Symplocaceae	Symplocos guianensis						1
Turneraceae	Turnera glaziovii						1
Urticaceae	Cecropia latiloba			1			
Violaceae	Gloeospermum sphaerocarpum		1				1
	Leonia glycicarpa		1				
	Paypayrola hulkiana		2				
	Rinorea					1	
	Rinorea amapensis		2				
	Rinorea pubiflora		1				
	Rinorea pubiflora var. pubiflora		1	1			
Vitaceae	Cissus erosa						1
	Cissus verticillata			2			
Zingiberaceae	Renealmia monosperma		1				

Grensgebergte granite mountain. The species has a South American distribution, and was also recorded in the Guianas in the states of Amazonas and Bolivar in Venezuela, Guyana and French Guiana (Funk et al. 2007, Discover Life 2013, GBIF 2013).

Platymiscium filipes was collected in Brasil and French Guiana, but so far not in Suriname (Funk et al. 2007, GBIF 2013, Tropicos 2013). This liana was collected in the 'seasonally flooded forest' along the boven Palumeu River during our journey to the Makrutu camp.

We found the tree species *Tabernaemontana angulata* in plot 3 at RAP site 4 (Kasikasima) in 'tall dryland tropical forest on laterite/granite hills'. The tree species has an Amazonian distribution occurring in Brazil, Colombia, and Venezuela (Discover Life 2013, WCSP 2013). The species has also been recorded from French Guiana and Guyana, but so far not in Suriname (Funk et al. 2007).

Prestonia cayennensis is the eighth newly recorded species for Suriname. This liana species has an Amazonian distribution, and was collected before in the Guianas in French Guiana, Guyana, and Venezuela (state of Bolivar), as well as in Brazil and Colombia but not in Suriname (Funk et al. 2007, Discover Life 2013). We found this liana in the seasonally flooded forest along the Palumeu River.

We found *Paloue induta* for the first time in Suriname along the edge of the Palumeu River in seasonally flooded forest close to RAP site 4 (Kasikasima). The tree species is known from Guyana and Brasil, but has so far to our knowledge, not been recorded for Suriname (Funk et al. 2007, GBIF 2013).

Heteropterys orinocensis is the tenth newly recorded species for Suriname. This liana species is known from Northern Brasil, and western Amazonia (e.g. Colombia, Venezuela, Peru), French Guiana, and from the states of Amazonas and Bolivar in Venezuela (Funk et al. 2007, GBIF 2013). It was found in the seasonally flooded forest along the Palumeu River at site 4.

We found the herb species *Ischnosiphon hirsutus* for the first time in Suriname in a wet patch of '(seasonally flooded) palm swamp forest and creek forest' at site 1. As far as we can assess the species has not been recorded for Suriname. *Ischnosiphon hirsutus* has been recorded in western Amazonian (e.g. Colombia, Peru, Bolivia, Colombia), the state of Bolivar in Venezuela, southern Guyana and Northwestern Brasil (Funk et al. 2007, GBIF 2013).

Clidemia attenuata is the twelfth newly recorded species for Suriname, and was found together with *Clusia flavida* at the edge of the tall dryland tropical forest on laterite/granite hills, on the Grensgebergte Mountain. The species was previously collected in French Guiana, Guyana and Venezuela (Funk et al. 2007, Tropicos 2013, GBIF 2013), but according to the data we have consulted, never before in Suriname.

The next newly recorded species for Suriname is *Clidemia epiphytica*. The species was previously recorded in western Amazonian, but was found in French Guiana as well (Funk et al. 2007, GBIF 2013, Tropicos 2013).

At RAP site 1 (Upper Palumeu) and site 4 (Kasikasima) we found the tree species **Quiina indigofera** in plots 1, 2 and 3 in 'tall dryland tropical forest on laterite/granite hills'. The species has been collected before in Brazil, Colombia, Guyana and Venezuela (states of Bolivar and Delta Amacuro), but according to the data we assessed, never in Suriname (Funk et al. 2007, GBIF 2013, Tropicos 2013).

The last and fifteenth species newly recorded for Suriname was **Micropholis splendens**. This tree species was found in plot 3 at RAP site 4 (Kasikasima) in 'tall dryland tropical forest on laterite/granite hills'. The species has been collected before in Brazil, Colombia, French Guiana, Guyana, and Venezuela (Funk et al. 2007, GBIF 2013, Tropicos 2013).

Noteworthy species

Several of the plant species we have collected during the expedition are interesting to mention in this report. This is mostly because of their rarity in geographical distribution and/or because the amount of known collections in the Guianas collections at National Herbarium of the Netherlands (NHN) is very small.

The palm tree *Syagrus stratincola* (Arecaceae) was collected in 'savannah forest on granite rock' on the top of the Kasikasima Mountains. *Syagrus stratincola* is a rare palm species endemic to the Guianas, and only known from ten localities in the Guianas (IUCN Redlist 2013). The status of this palm species on the IUCN Red List is Vulnerable. In Suriname this palm species is known from the Sipaliwini Savannah, the Tapanahony and Palumeu Rivers. The palm species occurs on granite plates and rocky outcrops. We also observed solitary trees of this palm in savannah forest on granite plates at the rapids along the Palumeu River and at the bathing place of camp 4, and in great abundance on the steep slopes of the Kasikasima Mountains.

We found the tree species *Ouratea cerebroidea* in plot 3 at RAP site 4 (Kasikasima) in 'tall dryland tropical forest on laterite/granite hills'. The species was previously recorded in Brazil, Guyana and French Guiana according to the encyclopedia of life (Funk et al. 2007, EOL 2013, GBIF 2013). The first collection of this species was collected by de Granville in the Tumuc Humac Mountains close to the border of French Guiana (GBIF 2013). We have collected the second collection for Suriname.

The orchid *Phragmipedium lindleyanum* found by us on top of the Grensgebergte RAP site 2 in open rock vegetation is a rarely collected species in Suriname (see page 26). According to Werkhoven (1986) and the National Herbarium of the Netherlands (NHN) collection, *Phragmipedium lindleyanum* is known in Suriname from the Wilhelminagebergte at an elevation above 800 m asl. It was collected by Stahel in 1926. The species most likely has a Guiana Shield distribution and has been recorded in Brazil, French Guyana, Guyana, and Venezuela and is associated with elevations above 500 m asl (WCSP 2013).

In the vicinity of plot 4 in a mixed forest of 'savannah forest on granite rock' and 'tall dryland tropical forest on laterite/granite hills', we found the tree species *Eugenia tetramera*, with large velvet yellow haired fleshy fruits. This tree belongs to a rare Myrtaceae species, and was also collected by O.S. Bánki and Frits van Troon in 2005 close to Stondansi in Northwestern Suriname. The current collection of *Eugenia tetramera* is the sixth collection in total for Suriname according to NHN. Apart from Suriname, *Eugenia tetramera* has been recorded in Brazil, French Guiana and Guyana, and the species has a Guianan distribution (Funk et al. 2007, GBIF 2013).

According to the NHN *Ixora piresii* (Rubiaceae) was collected for the last time in 1975 on the Lely Mountains, by Lindeman et al. It was a tree of 1.5 meters high and was collected in the understory of the tall dryland forest on Laterite Hill, at the Kasikasima Mountain. Specimens of this species were also collected on the Nassau Mountains and along the Boven Coppename River. The species is also known from Brazil and French Guiana (Funk et al. 2007, EOL 2013).

Along the Palumeu River at site 4 (Kasikasima) we found the noteworthy tree species *Taralea oppositifolia* in the tree layer of the seasonally flooded forest. *Taralea oppositifolia* was collected in Suriname in 1936 along the Litanie River for the last time. We collected the fourth specimen for Suriname and the first with fruits. *Taralea oppositifolia* has a neotropical distribution. It was recorded from western to eastern Amazonia, for example Colombia, Venezuela, Peru, Brazil, the Guianas and Panama, Dominican Republic and Peru, (Funk et al. 2007, GBIF 2013).

On top of the Grensgebergte RAP site 2 on the forest floor of a short mountain savannah forest we found *Lembocarpus amoenus* a Gesneriaceae herb with one leaf and an inflorescence stalk with purple bell shaped flowers. Our collection of *Lembocarpus amoenus* is the second collection for Suriname at the NHN. The first specimen was collected on the Bakhuis Mountain, in 1965. Apart from Suriname, *Lembocarpus amoenus* has been recorded in French Guiana. The species has a restricted distribution and is so far only found on ferro-bauxite Mountains (Funk et al. 2007, EOL 2013, GBIF 2013).

On the top of the Grensgebergte RAP site 2, on the Kasikasima Mountains, and on rocky outcrops close to large *Sinningia incarnata*. This species is another Gesneriaceae herb with showy red flowers. *Sinningia incarnata* typically occurs on granitic rocks and is the second collection for Suriname. The first specimen is known from the Voltzberg, collected by de Granville et al. in 1932 (GBIF 2013). *Sinningia incarnata* has a neotropical distribution (Funk et al. 2007, GBIF 2013).

In seasonally flooded palm swamp forest and creek forest at RAP site 4 (Kasikasima) we encountered *Costus lanceolatus* subsp. *pulchriflorus* (see page 26). This herb species with showy red flowers is the fourth collection for Suriname at the NHN. It was previously collected in Oelemari and Taponte (Marowijne River). *Costus lanceolatus* subsp. *pulchriflorus* has also been recorded in Brazil and French Guiana (Funk et al. 2007, GBIF 2013).

Cavendishia callista (Ericaceae) is the fifth collection for Suriname according to the NHN. It was collected on the plateau of the Grensgebergte Granite Mountain, in full bright sun. It was previously recorded on the Hendriktop (1922), Bakhuis Mountain (1965) and Lely Mountain (1975, 2004) in Suriname. *Cavendishia callista* has a North South American and Middle American distribution (GBIF 2013).

IUCN Red list species

We encountered three tree species listed on the IUCN Red List, namely:

- *Minquartia guianensis* Lower Risk (LR)/near threatened (NT) ver 2.3 (1994)
- *Syagrus stratincola* Vulnerable B1+2c ver. 2.3, 1998
- *Vouacapoua americana* Critically Endangered (CR) A1cd+2cd ver 2.3 (1994)

It needs to be noted that *Vouacapoua americana* within Suriname can still be found at some places in reasonable amount of numbers. In total, we found 9 individual trees of *Vouacapoua americana* spread over all the four 0.1 ha plots.

We encountered 9 species of orchids that are listed on the CITES appendices. Three orchid species still need to be identified. The other six orchids are *Cleistes rosea*, *Dichea picta*, *Epidendrum nocturnum* (see page 26), *E. densiflorum*, *Maxillaria discolor* and *Phragmipedium lindleyanum*.

We encountered one tree species, *Manilkara bidentata* (boletri) that is protected against felling by the Surinamese law (Surinamese Forest Law of 1992).

Plot inventories

In all four 0.1 ha plots combined we found a total of 791 individual trees belonging to approximately 44 families, 79 genera, and 139 (morpho-) species. Of the total 791 individual trees, 1.4 % could so far not be identified to the family level, 16 % could so far not be identified to the genus level, and 32% of the species could only be identified to morpho-species level so far. The five most common tree families were: Fabaceae (175 individuals), Burseraceae (67), Sapotaceae (60), Lecythidaceae (55), and Apocynaceae (45). The five most common tree genera in the four plots were: *Inga* (50 individuals), *Eperua* (42), *Swartzia* (32), *Siparuna* (31), and *Pouteria* (30). The ten most common tree species across the four plots were: *Eperua falcata* (42 individuals), *Siparuna cuspidata* (30), *Inga acrocephala* (28), *Tabernaemontana undulata* (27), *Micropholis guyanensis* (26), *Bocoa virdiflora* (25), *Tetragastris altissima* (21), *Tetragastris panamensis* (20), *Brosimum guianense* (19), and *Toulicia pulvinata* (19).

Although the mean Fisher's alpha of the plots from southern Suriname (the plots of the current study combined with those from the Kwamalasamutu surroundings) was higher (Fa = 41.82) than the mean Fisher's alpha of the plots from northern Suriname (Fa = 31.74), the difference was not significant. The plots in southern Suriname therefore did not have a significantly higher tree alpha diversity. The difference in total amount of species and total amount of individual trees per plot was also not significant between the plots of the southern and northern Suriname.

The Nonmetric Multidimensional Scaling (NMS) ordinations for all species, species with five or more individuals in total, and all genera showed the same patterns in floristic differences between plots. Most variation in floristic differences was explained by the difference between the plots from southern Suriname and plots from northern Suriname. There were however still some floristic differences between the plots of the Kwamalasamutu surroundings and the plots of this current study (see Figure 3.1 and 3.2).

The results of the Indicator Species Analyses showed that only a small number of species and genera were responsible for the floristic differences observed between the plots of southern and northern Suriname. Of all species in the total plot database, 11 % had five individuals or more. Of these species, we found 62 % of the species to be a significant indicator for either the plots in the northern Suriname or plots in southern Suriname (plots in the surroundings of Kwamalasamutu combined with plots of the current study). For the genera we found 12 % of all genera to be a significant indicator for either the plots of northern Suriname or plots in southern Suriname (plots in the surroundings of Kwamalasamutu combined with plots of the current study).

Within the four 0.1 ha plots combined of the current study we found 24% of the tree species to be listed as commercial timber species according to the Surinamese forest law of 1992. These 24% of all tree species in the plots made up 39% of the total individual trees found across all the four plots. In the plots of northern Suriname, 33% of all the tree species were listed as commercial timber species according to the Surinamese forest law of 1992. These 33% of the total tree species made up however 25% of the total amount of individual trees found in the six plots from northern Suriname. Although less commercial timber species seemed to occur in the four 0.1 ha plots of this study, these timber species did represent more individual trees in comparison to the plots of northern Suriname. It needs to be noted that quite common commercial tree species, like *Eperua falcata*, were represented by most individual trees. In the study in the surroundings of Kwamalasamutu, very few commercial timber species were observed. These findings showed the forests in the south of Suriname are not particularly rich in commercial timber species.

DISCUSSION

Of the total amount of 354 species found during the expedition to the regions of Grensgebergte and Kasikasima and identified so far, about 4% represent new records for Suriname. Eight of these newly recorded species have an

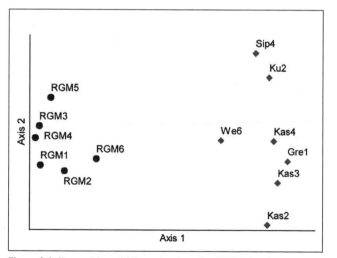

Figure 3.1. Non-metric multidimensional scaling (NMS) showing the floristic differences of the 0.1 ha plots of the South of Suriname, current study (Gre1, Kas2-4) combined with plot from the surroundings of Kwamalasamutu (Ku2, Si4, We6), and the 0.1 ha plots of Northern/Central Suriname (RGM 1-6). The NMS shown is the ordination based on all species with five or more individuals in the total plot data set of the South and North of Suriname. The first axis represented most variation in floristic differences (77%) separating the plots of the South of Suriname from the plots of the Northern/Central Suriname. The second axis represented 14% of the floristic variation, mostly showing the floristic differences between the plots of the current study and the plots from the surroundings of Kwamalasamutu.

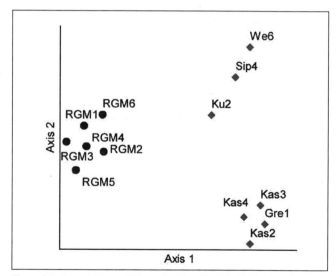

Figure 3.2. Non-metric multidimensional scaling (NMS) showing the floristic differences of the 0.1 ha plots of the South of Suriname, current study (Gre1, Kas2-4) combined with plot from the surroundings of Kwamalasamutu (Ku2, Si4, We6), and the 0.1 ha plots of Northern/Central Suriname (RGM 1-6). The NMS shown is the ordination based on all genera in the total plot data set of the South and North of Suriname. The first axis represented most variation in floristic differences (60%) separating the plots of the South of Suriname from the plots of the Northern/Central Suriname. The second axis represented 26% of the floristic variation, mostly showing the floristic differences between the plots of the current study and the plots from the surroundings of Kwamalasamutu.

Amazonian distribution, while six species have a Guianas or a Guiana Shield distribution, and one newly recorded species for Suriname has a Neotropical distribution. Of the total amount of 152 genera found during the expedition to the Grensgebergte and Kasikasima and identified so far, about 1% represent new records for Suriname. The genus *Tetrameranthus* has a Guianas distribution, while the genus *Tovomitopsis* has a Neotropical distribution. The geographical distribution of the new records for Suriname suggests that opposed to northern Suriname, southern Suriname could show more floristic affinity with the whole of Amazonia. Since many of the newly reported species are already known from other countries in the Guianas and the Guiana Shield, it could also show that the forests in the South of Suriname are little explored. We assessed the occurrence of the fifteen newly recorded species, and two genera from the best possible available digital information. It is however possible that valuable information about the occurrence of these species and two genera within Suriname has been missed, because the information is not digitally available. As several difficult plant families still require further identification, it is likely more new records or special species for Suriname will be found.

The new records for Suriname were found in four vegetation types, which are all dominant and floristically distinct for the South of Suriname. Five newly recorded species for Suriname were found in 'tall dryland tropical forest on laterite/granite hills', and four species on the edge of this forest type at the interchange to open rock vegetation. Five species were found in tall seasonally flooded forest along the Palumeu River, and one species was found in seasonally flooded palm swamp forest. Extensive areas of tall seasonally flooded forests were found along the Upper Palumeu River. The Upper Palumeu River is a meandering river that flows in valleys of a hilly landscape, and cuts through eroded sediment (loamy and sandy substrate). Because of the flat plains the water flows into the forests, just as was found in the surroundings of Kwamalasamutu (Bánki and Bhikhi 2011). Such extensive 'seasonally flooded forests' are less known from Northern parts of Suriname. Where the Upper Palumeu River starts to meander widely, vegetation types like palm swamps and 'tall herbaceous swamp vegetation and swamp wood' occurred frequently. These vegetation types resemble vegetation types in the coastal areas of Suriname, both in appearance and at times in species composition. Vegetation types like the 'tall dryland tropical forests on laterite/granite hills', 'savannah moss forest on granite rock' and 'open rock (Inselberg) vegetation, including rocky outcrops around rapids' are however dominant and exemplary for central (e.g. Raleigh Vallen) and southern Suriname. Between RAP Sites 1 and 4 there were some floristic differences in the composition of the 'tall dryland tropical forest on laterite/granite hills'. The open rock vegetation between the Grensgebergte (RAP Site 2) and the Kasikasima Mountains (RAP Site 4) had similar species composition. Interestingly, the rocky outcrops close to rapids along the

Upper Palumeu River also had comparable species composition to the open rock vegetation on the Grensgebergte and the Kasikasima Mountains. On these rock outcrops as well as on the Grensgebergte wet micro habitats with carnivorous plants were also similar. Such vegetation types are relatively rare. Most orchids that were found on the Grensgebergte rock outcrop were recorded only a few times in Suriname (Werkhoven 1986).

Previous studies predict that tree alpha diversity increases from the North of the Guianas to the South, especially in Guyana and Suriname (ter Steege et al. 2003, 2006). Our findings based on real data from seven 0.1 ha plots from southern Suriname show that tree alpha diversity is not significantly higher in comparison to northern Suriname. Most of the seven 0.1 ha plots, from the current study and from the study in the surroundings of Kwamalasamutu, occur in 'tall dryland tropical forest on laterite/granite hills'. This forest type is standing on relatively nutrient poor and shallow soils because of the bedrock underneath. It is likely the overall tree alpha diversity in this forest type is constrained by these ecological conditions. As the landscape in southern Suriname is dominated by laterite/granite hills, the forest type of 'tall dryland tropical forest on laterite/granite hills' is also very dominant in these areas. These results do not support a view that tree alpha diversity is higher in southern Suriname. However, it is possible that in some forest types tree alpha diversity is higher in the South in comparison to the Northern part of Suriname. The only plot in tall lowland rainforest standing on top of a well-drained loamy to sandy soil along the Kutari River has one of the highest tree alpha diversity recorded for Suriname so far (Bánki and Bhikhi 2011). It is possible that the tall lowland rainforest in the South of Suriname has a high tree alpha diversity in general in comparison to the Northern part of Suriname. But we currently lack the data to firmly state this.

The forests in South Suriname differ substantially in floristic composition from the forests in the Northern part of Suriname that also occur on the Guiana Shield basement complex. The majority of the floristic variation in the total plot database of 13 0.1 ha plots, showed a clear separation between the plots of the South of Suriname with those from the North of Suriname. All the non-metric multidimensional scaling (NMS) analyses showed similar patterns in floristic variation between the plots of the South of Suriname and plots of the Northern part of Suriname. Moreover, there are specific and significant indicator species and genera for the South and the North of Suriname, as revealed by the Indicator Species Analyses. These results indicate that the 'tall dryland tropical forests on laterite/granite hills' found in most of the plots in southern Suriname, have floristic elements that do not occur in the plots of northern Suriname. A lot of identifications of especially the sterile tree species collections still have to be resolved both for the current study and for the plots in the surroundings of Kwamalasamutu. When these tree identifications will be included, we expect that the floristic differences between the plots in the Kwamalasamutu surroundings and those of the current study will become smaller. Most likely, the completion of sterile tree collection identifications from the plots of the current study and for the plots in the surroundings of Kwamalasamutu will also lead to even sharper floristic differences between the forests of southern and northern Suriname. Our findings also do not support a view that the forests of the Guiana basement complex in Suriname all belong to one uniform forest type (see also Bánki and Bhikhi 2001). This is assumed by WWF's ecoregions, where most forests covering the Guianas are referred to as Guianan moist forest (see www.wwfguianas.org/about_guianas/ecoregions.cfm). Our findings rather support a view that the forests in southern Suriname have several unique floristic elements and are poorly explored. These floristic elements warrant protection of a wilderness area in the Southern Guianas as identified and proposed during the Guayana Shield Conservation Priority setting workshop in Suriname 2002 (Huber and Foster 2003).

CONSERVATION RECOMMENDATIONS

With regard to the plants and forest surveyed during the Grensgebergte and Kasikasima RAP expedition, some conservation recommendations can be made.

The Grensgebergte and the Kasikasima Mountains contain several vegetation types that are within Suriname dominant and floristically distinct for the central and southern parts of Suriname. Within these vegetation types, we recorded nearly all of the fifteen new plant species and two new genera for Suriname we found during this study. We also recorded several species with a restricted distribution in Suriname and/or in the Guianas, orchids listed on Appendices I and II of CITES, some carnivorous plants and three tree species that are listed on the IUCN Red List. Amongst these is a unique palm species *Syagrus stratincola* that is only know from ten localities in the Guianas, and is listed as Vulnerable on the IUCN Red List. We also found the tree species *Vouacapoua americana* that is listed as Critically Endangered on the IUCN Red List. As plant collections from several plant families still await identification, we expect to find more new records or noteworthy species for Suriname. These findings indicate the pristine status of the forests and vegetation types in southern Suriname, and the fact that these forests are still poorly explored.

Based on plot studies, we found distinct differences between the forests in the South from those examined in the North of Suriname. We also found some floristic differences between the forests in the South of Suriname of this study and from the rapid assessment in the surroundings of Kwamalasamutu. However, the floristic differences between plots in the South of Suriname were far less distinct, in comparison to the floristic differences between the plots of the South and the plots of the North of Suriname. Our findings contrasts current models in tree alpha diversity in the Guianas that predict tree alpha diversity ought to be higher

in southern Suriname. Species composition of the forests in the South of Suriname was found to be different from the North, and a small set of significant indicator species and genera were found for the South of Suriname. All the plots used in this chapter were located on the Guiana Shield basement complex. We therefore conclude that the forests on the Guiana Shield basement complex are not one uniform forest type as being suggested by some. Some forest types like the tall dryland forests on laterite/granite hills are more dominant in the South of Suriname, and are floristically distinct.

In combination with those from the Kwamalasamutu surroundings, the results of the plant rapid assessment in the surroundings of the Grensgebergte and the Kasikasima Mountains clearly show the occurrence of several dominant vegetation types with distinct floristic elements for Suriname. We expect that further exploration of the South of Suriname will only further strengthen these findings. The floristic distinctiveness of the South of Suriname warrants protection of a wilderness area in the Southern Guianas.

REFERENCES

Bánki, O.S. 2010. Does neutral theory explain community composition in the Guiana Shield forests? PhD dissertation, Universiteit Utrecht, Netherlands.

Bánki, O.S. and C.R. Bhikhi. 2011. Plant diversity and composition of the forests in the surroundings of Kwamalasamutu. *In:* O'Shea, B.J., L.E. Alonso and T.H. Larsen (eds.). 2011. A Rapid Biological Assessment of the Kwamalasamutu region, Southwestern Suriname. RAP Bulletin of Biological Assessment 63. Conservation International, Arlington, VA.

Discoverlife. 2013. www.discoverlife.org

Dufrene, M. and Legendre, P. 1997. Species assemblages and indicator species: the need for a flexible asymmetrical approach. Ecological Monographs, 67, 345–366.

Encyclopedia of Life (EOL). Available from http://www.eol.org. Accessed 3 February 2013.

Fisher, R.A., Corbet, A.S., and Williams, C.B. 1943. The relation between the number of species and the number of individuals in a random sample of an animal population. Journal of Animal Ecology, 12, 42–58.

Funk, V., Hollowell, T., Berry, P., Kellof, C., and Alexander, S.N. 2007. Checklist of the Plants of the Guiana Shield (Venezuela: Amazonas, Bolivar, Delta Amacuro; Guyana, Surinam, French Guiana). Contributions from the United States National Herbarium, 55, 584.

Global Biodiversity Information Facility (GBIF). Data accessed at the GBIF data portal at www.gbif.org accessed in February–March 2013.

Hall, A. 2012. Forests and Climate Change: The Social Dimensions of REDD in Latin America. Edward Elgar Pub.

Huber, O. and Foster, M.N. 2003. Nature conservation priorities for the Guiana Shield; 2002 consensus. Report and wall map. Conservation International, Netherlands Committee for IUCN and United Nations Development Programme, Washington.

IUCN. 2012. IUCN Red List of Threatened Species. Version 2012.2. www.iucnredlist.org. Downloaded on 26 March 2013.

Jansen-Jacobs, M.J. 1985–present. Flora of the Guianas. Royal Botanical Gardens Kew, Koelz Scientific Books.

Lindeman, J.C., and S.P. Moolenaar. 1959. Preliminary survey of the vegetation types of northern Suriname. Van Eeden Fonds, Amsterdam.

McCune, B. and Grace, J.B. 2002. Analysis of Ecological Communities. MjM Software Design, Gleneden Beach, Oregon, U.S.A.

McCune, B. and Mefford, M.J. 1999. PC-ORD. Multivariate Analysis of Ecological Data. MjM Software, Gleneden Beach, Oregon, U.S.A.

Prance, G.T. 2007. Flora da Reserva Ducke, Amazonas, Brasil: Chrysobalanaceae. In: Rodriguésia 58 (3): 493–531.

Surinamese forest law of 1992 (see www.sbbsur.org)

ter Steege, H., Pitman, N.C.A., Phillips, O.L., Chave, J., Sabatier, D., Duque, A., Molino, J.-F., Prevost, M.-F., Spichiger, R., Castellanos, H., von Hildebrand, P., and Vasquez, R. 2006. Continental-scale patterns of canopy tree composition and function across Amazonia. Nature, 443, 444–447.

ter Steege, H., Pitman , N.C.A., Sabatier, S., Castellanos, H., van der Hout, P., Daly, D.C., Silveira, M., Phillips, O., Vasquez, R. van Andel, T., Duivenvoorden, J., de Oliveira, A.A., Ek, R.C., Lilwah, R., Thomas, R.A., van Essen, J., Baider, C., Maas, P.J.M., Mori, S.A., Terborgh J., Nuñez-Vargas, P Mogollón, H. and Morawetz, W. (2003b). A spatial model of tree-diversity and -density for the Amazon Region. Biodiversity and Conservation 12: 2255–2276.

The plant list, a working list of all plant species 2013 (http://www.theplantlist.org/)

Tropicos. 2013. Tropicos.org. Missouri Botanical Garden. 26 Mar 2013. http://www.tropicos.org

WCSP. 2013. 'World Checklist of Selected Plant Families. Facilitated by the Royal Botanic Gardens, Kew. Published on the Internet; http://apps.kew.org/wcsp/

Werkhoven, M.C.M. 1986. Orchideeën van Suriname/Orchids of Suriname. VACO N.V. Uitgeversmaatschappij.

Westra, L.W.T, Maas P.J.M. 2012. *Tetrameranthus* (Annonaceae) revisited including a new species. PhytoKeys 12 : 1–21. doi: 10.3897/phytokeys.12.2771.

Chapter 4

Aquatic Beetles of the Grensgebergte and Kasikasima Regions, Suriname (Insecta: Coleoptera)

Andrew Short

SUMMARY

An extensive survey of aquatic beetles was conducted between 9–26 March 2012 in the Grensgeberge and Kasikasima regions of the Upper Palumeu River Basin, Suriname. More than 2500 specimens were collected representing 157 species in 70 genera. Twenty-six species and 8 genera are confirmed as new to science, with an additional 10–15 species likely being undescribed. Surprisingly, more species were recorded here than during the Kwamalasamutu Region RAP despite less collecting effort. Additionally, there was a high species turnover between these RAP sites: 40% of the species recorded here were not found in the Kwamalasamutu Region. The families Lutrochidae, Hydroscaphidae, and Torridincolidae are recorded from Suriname for the first time. While a broad range of habitats contributed to the high species and lineage diversity, hygropetric habitats on granite outcrops in particular provided a wealth of new and interesting taxa. Two of the new species in the family Hydrophilidae are described herein: *Tobochares kasikasima* Short sp. n. and *Tobochares striatus* Short sp. n. A key to the species of *Tobochares* is provided.

INTRODUCTION

Aquatic beetles—loosely defined as beetles that require aquatic habitats for at least one life stage—represent a significant portion of freshwater aquatic macroinvertebrate diversity with approximately 13,000 described species found worldwide (Jäch & Balke 2008). These species are distributed across approximately 20 beetle families in four primary lineages: Myxophaga, Hydradephaga, aquatic Staphyliniformia (Hydrophiloidea and Hydraenidae) and the Dryopoidae (or aquatic byrroids). Members of Myxophaga are small beetles that feed largely on algae as larvae and adults. The Hydradephaga (including the diving and whirligig beetles) are largely predators as adults and larvae; the aquatic Staphyliniformia are largely predators as larvae but scavengers as adults; the dryopoids are largely scavengers or eat algae as both larvae and adults.

Aquatic insects in general (including several groups of aquatic beetles) are often used to assess water quality in freshwater rivers and streams. The dryopoids are most frequently used for this purpose because they are most commonly found in these habitats and often have high-oxygen needs. Aquatic beetle communities are also effectively used to discriminate among different types of aquatic habitat (e.g. between lotic and lentic; rock outcrops, substrate, etc.).

The only prior survey to have focused on aquatic beetles in Suriname was a prior RAP survey in the Kwamalasamutu Region (Short and Kadosoe 2011). Overall, the fauna of the country and broader region is very poorly known, with the exception of recent work in Venezuela where the number of known species has literally more than doubled in just a few years.

METHODS AND STUDY SITES

I collected aquatic beetles at three of the four primary sites on the RAP (Site 1: Upper Palumeu; Site 2: Grensgebergte; Site 4: Kasikasima).

Field methods

I used a variety of passive and active collecting techniques to assemble as complete a picture of the aquatic beetle communities as possible. Passive techniques are advantageous because they often allow large amounts of material to be collected in quantitative ways at one time and with little effort, but provide little ecological or habitat data; we do not gain

The RAP Bulletin of Biological Assessment meets all criteria for the description of species new to science, as specified by the International Commission on Zoological Nomenclature (ICZN). Paper copies of the RAP Bulletin are deposited in the library of Conservation International, Arlington, VA, USA; the University of Kansas, Lawrence, KS, USA; and the National Zoological Collection of Suriname, Paramaribo, Suriname. The print run of this issue of the RAP Bulletin consisted of 300 copies.
— LEA, THL, August 2013

new insights into the water quality requirements of insects collected in this manner. In contrast, active collecting methods (i.e. by hand) provide a richer source of information on the microhabitat and water quality requirements of species, but are more time intensive, qualitative, and may suffer from collector bias.

Traps and other passive methods. During one night at Site 1, and four nights at Site 4, I collected in the evening hours until approximately 11 p.m. at a UV light mounted on a white sheet erected on the periphery of the camp. I also used flight intercept traps (FITs) to sample the beetle fauna. These traps collect flying insects, including dispersing aquatic beetles. At site 1, three FITs were used, each composed of a 2-meter wide by 1.5- meter high screen, with aluminum pans filled with soapy water as a collecting trough. No FITs were constructed by me at 2 or 4, but residues in traps at these sites by T. Larsen were saved and picked for aquatic beetles. Dung trap sample residues collected by T. Larsen were also were also examined for specimens of Hydrophilidae.

Active methods. For collection of actively swimming insects, I used a large aquatic insect net to survey forest pools, swamps, and the margins of rivers and streams. To collect poorly swimming insects that float to the water's surface when disturbed, I employed a small metal strainer to collect in detrital micropools and marginal areas. At Sites 1, 2, and 4, I collected in hygropetric habitats (waterfalls and rock seepages). This was done by "washing" the wet rock surface with a scrub brush into a small net, and examining the algae and debris for beetles. I also submerged moist leaf packs in a tub of water and collected the insects that floated to the surface.

Because most aquatic beetles are very small (generally <5 mm, although a select few reach 50 mm) and many require examination under a microscope for species identification, I collected and preserved samples of these insects from each camp to take back to the laboratory for processing.

Site 1: Upper Palumeu, (Juuru Camp) N 2.47700°, W 55.62941°, 277 m, 9–18 March 2012.
Most collecting was done along a cut trail south that continued to the Brazilian border, and in the immediate (<1 km) vicinity of basecamp. Along the trail a few kilometers south

Figure 4.1. Selected collecting event photographs, with field numbers given in brackets. A) Camp 1, detrital pool along floodplain of small upper Palumeu tributary [SR12-0311-01A], B) Camp 1, tributary of the upper Palumeu [SR12-0314-01A], C) Camp 2, typical seepage habitat on the Grensgebergte summit [SR12-0312-01A], D) Camp 4, large flowing seepage near the base of Kasikasima [SR12-0324-01C].

of basecamp was a large cascade on a tributary of the Palumeu River with large expanses of wet rock on the border of the river channel. The stream on which the waterfall was located has a mostly sand and detritus substrate (Figure 4.1B). Some forested low-lying areas along both this stream and the Palumeu held water when the survey began, forming shallow detrital pools (Figure 4.1A). Following rains and flooding of the Palumeu River, these areas became completely submerged and united with the main river as the water level rose approximately 2 meters in less than three days. The heavy rains and subsequent flooding of productive habitats reduced the overall collecting done at this site. Approximately three days of accumulated specimens were lost from two of the three FIT traps, as the traps were submerged by rising floodwaters.

Habitats of note: Sandbars and leaf packs along the above-mentioned Palumeu tributary proved very fruitful for some less common taxa (e.g. *Laccodytes*, *Hydrodessus*) until the rivers flooded. Some large *Megadytes* specimens were collected directly from the margin of Palumeu River itself, and could be seen by headlamp swimming at night. Hygropetric habitats at the cascade yielded unique taxa, including *Oocyclus* and some other rare water scavenger beetles.

Site 2: Grensgebergte, N 2.46554°, W 55.77034°, 790–820 m, 12–13 March 2012.

All collecting was done on a variety of seepage habitats on the granite rock (Figure 4.1C). These hygropetric areas were expansive in some areas, covering hundreds of square meters. While the entire rock becomes wet and slippery following

rains, certain "drainage" areas (channels in the rock and areas below water-absorbing vegetation) form more "perennial" seepages, although these are still likely seasonal. These more permanent seepage areas had large amounts of filamentous algae and liverworts in addition to the cyanobacterial films and lichens that were nearly ever-present on the outcrop.

With the exception of a few small depressions that held water only during rains, no streams, forest pools, or other aquatic habitats were observed along the cut transect or its vicinity. I did not descend from the ridge, although there are surely streams nearby.

Habitats of note: As described above, the site contained enormous swaths of hygropetric habitats, and while not rich in total number of species, many were rare or new to science.

Site 4: Kasikasima, N 2.97731°, W 55.38500°, 201 m, 19–26 March 2012.

Collecting at the Kasikasima site was divided into two primary areas: the lowland basecamp (c. 200 m) situated on the Palumeu River, and the Kasikasima rock formation itself (with collections made at 400 and 515 m). In the lowland area, I collected along trails between basecamp and a METS tourist lodge, and between basecamp and Kasikasima. Both trails crossed several small streams (c. 1–4 m in width), and had various combinations of mud, detritus, and/or sand as substrate. One stream originated at the base of Kasikasima and flowed over several collections of large boulders, creating a small area of hygropetric habitat. I also collected along the banks of the Palumeu River itself, as well as some flooded forest areas along its margin.

Habitats of note: At the lowland basecamp, there were no particularly unusual habitats. As is typical, the smaller, sandier streams yielded the most interesting and less commonly collected taxa. Along the trail ascending Kasikasima, there was a broad seepage at 400 m with moderate, perhaps semi-permanent, flow (Figure 4.1D; 2°58.613'N, 55°24.683'W), such that it was sustained even days after a rain event and also created a small stream channel when it left the rock. It appears this seepage captures and drains rainwater from a broad expanse of rock. This seepage was densely populated with many new and "rare" taxa, all restricted to hygropetric habitats. At the terminus of the trail (at a low summit of 515 m), there were also several localized seepage habitats, although these were ephemeral, and only present during and shortly after rain events; Nevertheless, several *Fontidessus* were collected here.

Sample & Specimen Processing

I made 27 separate collections over the course of the expedition. Specimens collected by hand and bulk trap samples were all placed directly into 95% ethanol. Following field-work, all samples were sorted, mounted, labeled and data-based. In some cases where there were long series common species from a single sample, only a portion was prepared and the remainder was stored and archived in ethanol. Only mounted specimens are reflected in specimen counts given

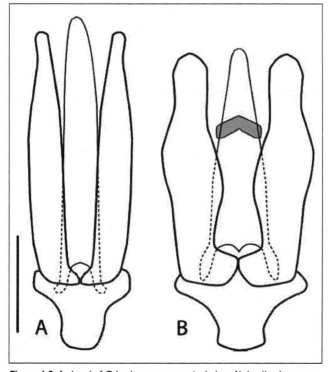

Figure 4.2. Aedeagi of *Tobochares* spp., ventral view. A) *kasikasima* sp. n., B) *striatus* sp. n. Scale bar = 0.1 mm.

in this report, thus abundances of the most common taxa are underreported. The morphospecies numbering system used in the Kwamalasamutu RAP (Short and Kadosoe 2011) is employed here so that the taxonomic diversity between expeditions can be directly compared (e.g. *Copelatus* sp. 3 refers to the same taxon in both reports).

RESULTS AND DISCUSSION

In total, 157 species in 70 genera of aquatic beetles were collected from the three RAP camps (Table 4.1; Figure 4.3). While the two primary camps (Upper Palumeu and Kasikasima) had similar numbers of taxa (92 and 105 species respectively) they were surprisingly dissimilar in composition, with each possessing ca. 50% site-unique species. Only 10 species in 8 genera were recorded from the Grensgebergte camp, but this was due to the lack of habitat diversity on the ridge—only hygropetric seepages were present, no trapping was conducted, and only limited time (2 days) were spent at the site collecting.

There were almost 10% more species collected during this RAP than the previous one around Kwamalasamutu, on which 144 species were recorded (Short and Kadosoe 2011). This is striking because far fewer overall specimens were collected (c. 2500 vs. >4000), in large part because of reduced manpower (one collector vs. two for Kwamalasamutu). Amazingly, more than 40% of the species (69) collected during this RAP were not collected at Kwamalasamutu. Some species that were extremely abundant in Kwamalasamutu (e.g. *Enochrus* sp. 1, *Notomicrus*, *Thermonectus*) were generally rare on this RAP. In some cases, such as with the majority of seepage taxa, this can be attributed to availability of suitable habitat. In other cases, it may be due to seasonality

differences (October vs. March). For most, however, there is no clear answer.

By far the most significant result of the expedition was the discovery of large Myxophagan communities at sites 2 and 4. This represents the first known collection of the families Hydroscaphidae and Torridincolidae in Suriname. Several species of Torridincolidae were abundant at both sites; comparatively few specimens of an unknown number of Hydroscaphidae taxa were present only at site 2. At least three genera of Torridincolidae were recognized, all undescribed, but also known to me from similar habitats in western Venezuela. Due to their small size (under 2 mm), further characterization of these exciting groups must wait detailed examination in the laboratory.

Table 4.1. Summary of aquatic beetle diversity among sites.

	# Specimens	# Genera	# Species	# Site-Unique species
Camp 1/ Upper Palumeu	1065	50	92	46
Camp 2/ Grensgebergte	155	8	10	3
Camp 4/ Kasikasima	1336	53	105	53
TOTAL	2556	70	157	-

Taxa of Note

Hydroscaphidae & Torridincolidae: This is the first report of either family and the suborder Myxophaga from Suriname (they are not yet known from either Guyana or French Guiana). With regard to Hydroscaphidae, a few individuals of a new species of *Scaphydra* were collected on an algae-covered seepage on the Grensgebergte rock summit. This is a typical habitat for hydroscaphids, and the family can be common in similar granite outcrops on the western edge of the Guiana Shield region in Venezuela (Short et al. 2010, unpub. data). Numerous new species of Torridincolidae (most in undescribed genera) were found on both the Grensgebergte summit as well as on a larger seep on the side of Kasikasima. As with Hydroscaphidae, this is a typical habitat for the family. While the genera are new, they are also known (but unpublished) from a variety of localities in southern Venezuela, also on granite seepages.

Hydrophilidae gen. nov. 3, sp. 1: This undescribed genus and species was also found on both the Grensgebergte summit and seep on the side of Kasikasima. As with the Myxophaga, the genus is known (but unpublished) from southern Venezuela, however, the Surinamese taxon reported here is confirmed as distinct from those species.

Lutrochus sp.: This is the first report of the family Lutrochidae from Suriname (Maier and Short 2013), and the taxon collected represents a new species.

Figure 4.3. Selected habitus images of aquatic beetle taxa collected in the Grensgebergte and Kasikasima regions. A) *Copelatus* sp., B) *Vatellus grandis*, C) *Derallus* sp., D.) *Anodocheilus* sp., E) *Platynectes* sp., F) *Dryops* sp.

Oocyclus spp.: Three species of this genus were found, all of which are already described. All three records represent significant eastern range extensions, as all three were previously known only from Venezuela (Short and Garcia 2010). Interstingly, *Oocyclus trio* was not encountered, which was described from the Kwamalasamuto RAP.

DESCRIPTION OF NEW SPECIES

Genus *Tobochares* Short & Garcia, 2007
The genus *Tobochares* currently contains two species known only from a handful of sites in the Guiana Shield region of Venezuela and Suriname. Two additional new species were discovered during the RAP. Like the previously described species, both new taxa are also apparently restricted to hygropetric habitats.

Tobochares kasikasima Short, sp.n.
Type Material: Holotype (male): "SURINAME: Sipaliwini District/ N 2°58.613', W 55°24.683', 400m/ Camp 4 (high) Kasikasima/ leg. A. Short; main seepage area/ 24.iii.2012; SR12-0324-01C/ 2012 CI-RAP Survey", "[barcode]/ SEMC1088588/KUNHM-ENT". (Deposited in the National Zoological Collection of Suriname).

Diagnosis. Elytra not sulcate (sulcate at least in basal half of all other known species) Maxillary palps uniformly yellow (darkened at apex in *T. sulcatus* and *T. striatus*). Aedeagus with parameres straight along inner and outer margins (sinuate in other all other known species).

Description. *Size and form.* Body length = 1.7 mm. Body elongate oval, weakly dorsoventrally compressed. *Color and punctation.* Dorsum of head very dark brown, almost black; anterolateral margins of clypeus without paler preocular patches; maxillary palps yellow, with apical palpomere slightly darkened at apex. Pronotum and elytra brown, becoming slightly paler at lateral margins. Meso- and metathoracic ventrites and visible abdominal ventrites dark brown, with prosternum, epipleura, and legs distinctly paler. Ground punctation on head, pronotum and elytra moderately fine. *Head.* Antennae with scape and pedicel subequal in length, and their combined length subequal to antennomeres 3–8. Maxillary palps with palpomeres 2 and 4 subequal in length with palpomere 3 slightly shorter. *Thorax.* Elytra with indistinct rows of serial punctures which are subequal to slightly larger than the ground punctures; most rows feebly impressed into grooves in posterior quarter. Elevation of mesoventrite forming a lateral carina which is raised into a very low, bunt tooth. Metaventrite with distinct median narrowly ovoid glabrous area that three-quarters as long as the length of the metaventrite. *Abdomen.* Abdominal ventrites uniformly and very densely pubescent. Apex of fifth ventrite evenly rounded. Aedeagus (Figure 4.2A) with basal piece very short, less than one-third as long as parameres. Parameres subparallel in basal two thirds, tapering to a blunt tip in anterior. Median lobe long, parallel sided and tapering at the tip; distinctly protruding beyond the apex of the parameres; gonopore not apparent.

Habitat. The single known specimen was collected on a flowing seepage over granite.

Etymology. Named for the inselberg on which it was found, Kasikasima in southwestern Suriname.

Tobochares striatus Short, sp.n.
Type Material: Holotype (male): "SURINAME: Sipaliwini District/ N 2.24554°, W 55.77000°, 800m/ Camp 2 (Grensgebergte Rock/ leg. A. Short; rock seepages/ 12.iii.2012; SR12-0312-01A/ 2012 CI-RAP Survey". (Deposited in the National Zoological Collection of Suriname). **Paratypes (11): SURINAME: Sipaliwini District:** Same data as holotype (3 exs.); Camp 1, Upper Palumeu, 10.iii.2012, leg. A. Short, small forest pool by boulders, SR12-0310-02A (1 ex.); Camp 4 (Kasikasima), 24.iii.2012, leg. A. Short, main seepage area, SR12-0312-01C (7 exs.). Paratypes are distributed between the National Zoological Collection of Suriname and the University of Kansas.

Diagnosis. Elytra sulcate along entire length (only sulcate in basal half in *T. sipaliwini*, and not sulcate in *T. kasikasima*). Maxillary palps darkened at apex (uniformly pale in *T. sipaliwini* and *T. kasikasima*). Aedeagus sinuate along inner and outer margins (straight in T. *kasikasima*). Size smaller (larger in *T. sulcatus*).

Description. *Size and form.* Body length = 1.9–2.0 mm. Body elongate oval, moderately dorsoventrally compressed. *Color and punctation.* Dorsum of head very dark brown, almost black; anterolateral margins of clypeus with distinctly paler preocular patches; maxillary palps yellow, with apical palpomere slightly darkened at apex. Meso- and metathoracic ventrites and visible abdominal ventrites dark brown, with prosternum, epipleura, and legs slightly paler. Ground punctation on head, pronotum and elytra very fine. *Head.* Antennae with scape and pedicel subequal in length, and their combined length subequal to antennomeres 3–8. Maxillary palps with palpomeres 2 and 4 subequal in length with palpomere 3 slightly shorter. *Thorax.* Elytra with ten rows of serial punctures which are strongly depressed into grooves along entire elytral length; serial punctures very coarse and distinct, 4–5 time large than ground punctures. Elevation of mesoventrite forming a lateral carina which is raised into a low tooth elevated to the same plane as the ventral surface of mesocoxae. Metaventrite with distinct median ovoid glabrous area that is nearly two-thirds as long as the metaventrite length. *Abdomen.* Abdominal ventrites uniformly and very densely pubescent. Apex of fifth ventrite evenly rounded. Aedeagus (Figure 4.2B) with basal piece short, ca. one-third as long as parameres. Parameres strongly sinuate on both inner and outer margins in apical half; median lobe extended past the apex of parameres; gonopore situated distinctly below apex of the dorsal strut.

Habitat. Most specimens were collected on a flowing seepage on granite. A single specimen was collected in a small forest pool near Camp 1 on the upper Palameu River, although this pool was situated directly beneath a group of large granite boulders.

Etymology. Striatus, in reference to the grooved elytra.

Key to the Species of *Tobochares*

1 Elytra sulcate along entire length...2
- Elytra sulcate on posterior half or less, or almost appearing without grooves...3

2 Size larger (2.2–2.5 mm). (Venezuela)...*sulcatus* Short & Garcia
- Size smaller (1.9–2.0 mm). (Suriname)...*striatus* Short sp. n.

3 Apical maxillary palpomere uniformly yellow, not darkened at apex. Clypeus with preocular pale spots. Size larger (1.8–2.0 mm). Elytra sulcate on posterior half. (Suriname)...*sipaliwini* Short & Kadosoe
- Apical maxillary palpomere with apex darkened. Head uniformly dark brown to black. Size smaller (1.7 mm). Elytra only feebly sulcate on posterior quarter to fifth. (Suriname)...*kasikasima* Short sp. n.

CONSERVATION ISSUES & RECOMMENDATIONS

The water beetle diversity found at each camp was reasonable given the types of aquatic habitat present. The high number of genera and species, which cover a variety of ecological and habitat types, suggest the area is largely undisturbed. Some of the difference in species diversity between Camps 1 and 4 are likely due to stochastic sampling biases, although presence of some taxa at Camp 1 but not 4 (e.g. *Cybister* and *Berosus)* is unexplained. The large degree of site-specific taxa between Camp 1 and 4 (ca. 50% for each site) is rather exceptional, and cannot be explained by the presence of more seepage habitats at Camp 4 alone. No differences in the water beetle communities between the sites could be attributed to anthropogenic disturbance. The small number of species at Site 2 is attributable to the very reduced number of habitats (e.g. only rock seepages).

The hitherto virtually unknown diverse hygropetric fauna found at Camps 2 and 4, including several new family records, underscores how little we know about this area of Suriname. More work is necessary in other similar comparable habitats in Suriname and the region to determine how endemic these seepage taxa may be, and to better understand the full diversity of this evolutionarily and ecologically significant group. Between both RAP expeditions, 213 species of aquatic beetles were collected...more than had been recorded from all of the Guianas combined in the last 250 years. The high turnover between this expedition and the Kwamalasamutu RAP survey, combined with the high turnover between camps in a single RAP survey, suggest a highly diverse and very localized or patchily distributed water beetle community in southern Suriname. The findings of dozens of species new

to science on both RAP surveys, while not unexpected given the paucity of collecting in the region, reinforces how much further we have to go before we have an understanding of the biological diversity of southern Suriname.

ACKNOWLEDGMENTS

Special thanks are due to Vanessa Kadosoe (University of Suriname) for assistance in sorting and identifying material. Kelly Miller (University of New Mexico) and Crystal Maier (University of Kansas) provided helpful assistance with identifications of the Dytiscidae and Dryopoidea respectively. Sarah Schmits provided key databasing assistance and support.

REFERENCES

Jäch, M. and M. Balke. 2008. Global diversity of water beetles (Coleoptera) in freshwater. Hydrobiologia. 595: 419–442.

Maier, C. and A.E.Z. Short. 2013. A revision of the Lutrochidae (Coleoptera) of Venezuela with description of six new species. Zootaxa. 3637: 285–307.

Short, A.E.Z. and M. García. 2007. *Tobochares sulcatus*, a new genus and species of water scavenger beetle from Amazonas State, Venezuela (Coleoptera: Hydrophilidae). Aquatic Insects. 29: 1–7.

Short, A.E.Z. and M. Garcia. 2010. A review of the *Oocyclus* Sharp of Venezuela with description of twelve new species (Coleoptera: Hydrophilidae: Laccobiini). Zootaxa. 2635: 1–31.

Short, A.E.Z. and V. Kadosoe. 2011. Aquatic beetles of the Kwamalasamutu region, Suriname (Insecta: Coleoptera). *In:* O'Shea, B.J., L.E. Alonso and T.H. Larsen (eds.) 2011. A Rapid Biological Assessment of the Kwamalasamutu region, Southwestern Suriname. RAP Bulletin of Biological Assessment 63. Conservation International, Arlington, VA.

Appendix 4.1. Water beetles collected during the 2012 RAP survey of Southeastern Suriname

Taxon	Upper Palumeu	Grensgebergte	Kasikasima
DYTISCIDAE			
Agaporomorphus sp. 1	-	-	X
Amarodytes sp. 3	X	-	-
Anodocheilus sp. 1	-	-	X
Anodocheilus sp. 2	X	-	-
Bidessodes charaxinus Young, 1986	X	-	X
Bidessodes semistriatus Regimbart, 1901	X	-	X
Bidessodes sp. 3	X	-	-
Bidessonotus sp. 1	X	-	-
Celina sp. 2	-	-	X
Copelatus geayi Regimbart, 1904	X	-	X
Copelatus sp. 2	X	-	X
Copelatus sp. 3	X	-	X
Copelatus sp. 5	X	-	X
Copelatus sp. 6	X	-	-
Copelatus sp. 7	X	-	X
Copelatus sp. 8	-	-	X
Copelatus sp. 9	X	-	X
Copelatus sp. 10	X	-	X
Copelatus sp. 11	X	-	X
Copelatus sp. 12	X	-	-
Copelatus sp. 17	X	-	X
Copelatus sp. 18	X	-	X
Desmopachria sp. 1	-	-	X
Desmopachria sp. 3	X	-	-
Desmopachria sp. 4	-	-	X
Desmopachria sp. 5	-	-	X
Desmopachria sp. 8	-	-	X
Desmopachria sp. 9	-	-	X
Desmopachria sp. 10	X	-	X
Desmopachria sp. X	X	-	X
Fontidessus sp. 1*	-		X
Fontidessus sp. 2*	-	X	X
Hydaticus fasciatus LaPorte, 1835	X	-	-
Hydaticus sp. 2	X	-	-
Hydrodessus sp. 1	X	-	-
Hydrodessus sp. 4	X	-	-
Hydrodessus sp. 5	X	-	-
Hydrodytes sp. 1	-	-	X
Hypodessus sp. 1	X	-	X
Hypodessus sp. 2	X	-	-
Laccodytes apalodes Guignot, 1955	X	-	-

table continued on next page

Appendix 4.1. continued

Taxon	Upper Palumeu	Grensgebergte	Kasikasima
Laccodytes sp. 4	X	-	-
Laccophilus sp. 1	X	-	-
Laccophilus sp. 2	X	-	-
Laccophilus sp. 3	X	-	X
Laccophilus sp. 5	X	-	X
Laccophilus sp. 6	X	-	X
Laccophilus sp. 7	-	-	X
Laccophilus sp. 8	-	-	X
Laccophilus sp. 9	X	-	-
Laccophilus sp. 10	X	-	X
Laccophilus sp. 11	-	-	X
Megadytes sp. 1	X	-	-
Megadytes sp. 2	X	-	-
Microdessus sp. 1	-	-	X
Neobidessus spangleri Young, 1977	-	-	X
Pachydrus sp. 2	-	-	X
Platynectes sp. 1	-	-	X
Platynectes sp. 2	-	-	X
Thermonectus sp. 2	X	-	-
Vatellus grandis Buquet, 1840 [=sp. 2]	X	-	X
Vatellus tarsatus (LaPorte, 1835)	X	-	-
DRYOPIDAE			
Dryops sp. 1	X	-	X
Pelonomus sp. 1	-	-	X
Pelonomus sp. 2	X	-	-
ELMIDAE			
Cylleopus sp. 1	X	-	-
New Genus 1, sp. 1	X	-	-
Hintonelmis	X	-	-
Macrelmis sp. 1	X	-	-
Neoelmis sp. 1	X	-	-
Phaenocerus sp. 1	X	-	-
Phaenocerus sp. 2	-	-	X
EPIMETOPIDAE			
Epimetopus sp. 3*	-	-	X
GYRINIDAE			
Gyretes sp. 1	X	-	X
Gyretes sp. 2	-	-	X
Gyretes sp. 3	-	-	X

table continued on next page

Appendix 4.1. continued

Taxon	Upper Palumeu	Grensgebergte	Kasikasima
HYDRAENIDAE			
Hydraena sp. A	X	-	-
Hydraena sp. B	X	-	-
Hydraena sp. C	X	-	-
HYDROCHIDAE			
Hydrochus sp. 1	-	-	X
Hydrochus sp. 3	-	-	X
Hydrochus sp. 6	X	-	-
Hydrochus sp. 7	X	-	-
HYDROPHILIDAE			
Anacaena cf. *suturalis*	-	-	X
Anacaena sp. 2	X	-	-
Anacaena sp. 3	-	-	X
Australocyon sp. 1	X		X
Australocyon sp. 2	X	-	X
Berosus sp. 1	X	-	-
Cercyon sp. 2	X	-	-
Cercyon sp. 3	-	-	X
Cercyon sp. 4	X	-	-
Cercyon sp. 5	-	-	X
Cercyon sp. 6	-	-	X
Cercyon sp. 7	X	-	X
Cetiocyon incantatus Fikacek & Short, 2010	X	-	X
Chasmogenus sp. X*	X	-	X
Cyclotypus sp. 1*	X	-	-
Derallus intermedius Oliva, 1995	X	-	X
Derallus perpunctatus Oliva, 1983	-	-	X
Derallus sp. 2	-	-	X
Derallus sp. 3	-		X
Enochrus sp. 1*	X	-	X
Enochrus sp. 4	-	-	X
Enochrus sp. 6	-	-	X
Enochrus sp. 7	-	-	X
Globulosis sp. 1*	X		X
Guyanobius sp. 1	-		X
Helochares sp. 1*	X	-	-
Helochares sp. 2	X	-	X
Helochares sp. 4	-	-	X
Helochares sp. 6*	-	-	X
Hydrobiomorpha sp. 1	X	-	-
Moraphilus sp. 1	X	-	X

table continued on next page

Appendix 4.1. continued

Taxon	Upper Palumeu	Grensgebergte	Kasikasima
Gen. Nov. 1, sp 1*	-	-	X
Gen. Nov. 1, sp 2*	X	-	X
Gen. Nov. 2, sp. 1*	-	-	X
Gen. Nov. 3, sp. 1*	-	X	X
Gen. Nov. 4, sp. 1*	-	X	-
Gen. Nov. 4, sp. 2*	-	X	X
Notionotus shorti Queney, 2010	X	-	-
Notionotus sp. 2*	-	-	X
Notionotus sp. 3*	X	-	-
Oocyclus coromoto Short & Garcia, 2010	-	X	X
Oocyclus floccus Short & Garcia, 2010	X	-	X
Oocyclus petra Short & Garcia, 2010	-	X	-
Oosternum sp. 2	X	-	X
Oosternum sp. 3	X	-	X
Oosternum sp. 4	X	-	X
Pelosoma sp. 1	-	-	X
Pelosoma sp. 3	X	-	-
Pelosoma sp. 4	X	-	-
Pelosoma sp. 5	X	-	X
Pelosoma sp. 6	-	-	X
Pelosoma sp. 7	X	-	-
Phaenonotum sp. 3	-	-	X
Phaenostoma sp. 2	X	-	-
Phaenostoma sp. 4	-	-	X
Quadriops sp. 1	X	-	-
Quadriops sp. 2	-	-	X
nr. *Quadriops* sp. 1*	X	-	-
nr. *Quadriops* sp. 2*	-	-	X
Sacosternum sp. 1	-	-	X
Tobochares sipaliwini Short & Kadosoe, 2011	-	-	X
Tobochares striatus sp.n.	X	X	X
Tobochares kasikasima sp.n.	-	-	X
Tropisternus chalybeus LaPort, 1840	X	-	X
Tropisternus setiger Germar, 1824	X	-	-
HYDROSCAPHIDAE			
Scaphydrus sp. 1*	-	X	-
NOTERIDAE			
Notomicrus sp. 1	-	-	X
Notomicrus sp. X	X	-	X
Siolius bicolor Balfour-Browne, 1969	-	-	X
Suphisellus sp. 1	X	-	X

table continued on next page

Appendix 4.1. continued

Taxon	Upper Palumeu	Grensgebergte	Kasikasima
LUTROCHIDAE			
Lutrochus sp. 1*	-	-	X
TORRIDINCOLIDAE			
Gen. Nov. 1, sp. 1*	-	X	X
Gen. Nov. 2, sp. 2*	-	X	X
Iaper sp. 1*	-	-	X
TOTAL:	92	10	105

[*=new species]

Chapter 5

Dung beetles of the Upper Palumeu River Watershed (Grensgebergte and Kasikasima) of Southeastern Suriname (Coleoptera: Scarabaeidae: Scarabaeinae)

Trond H. Larsen

SUMMARY

Dung beetles are among the most cost-effective of all animal taxa for assessing biodiversity patterns, yet RAP's recent surveys are among the few that are expanding our knowledge of Suriname's little known dung beetle fauna. In addition to cost-effective sampling using standardized pitfall traps, dung beetles depend upon large mammals for food and consequently can be used to rapidly assess the health of the overall mammal community and hunting impacts in a fraction of the time it would take to survey the mammals themselves. I sampled dung beetles using baited pitfall traps and flight intercept traps in the Grensgebergte and Kasikasima regions of Southeastern Suriname. I collected 4,483 individuals represented by 107 species. This ranks among the most diverse places on the planet for dung beetles, and exceeds the extraordinarily high species richness observed in nearby southwestern Suriname (94 species, Larsen 2011). Ten species are most likely new to science, while an additional 10–20 species may be undescribed pending further taxonomic revisions.

Dung beetle species richness, abundance and biomass were higher around Upper Palumeu than at Kasikasima, probably due to the extensive intact forest and lack of hunting pressure in this remote headwater region where no one currently lives. Dung beetle diversity and abundance at Kasikasima were still relatively high, indicating only mild to moderate hunting of large mammals and birds in the region. All sites, including the Grensgebergte Mountains, supported high endemism, including several rare species, demonstrating the exceptional biodiversity value of the region. Surprisingly, dung beetle species composition varied strongly among sites within this survey, as well as among sites sampled on previous surveys, including nearby southwestern Suriname. This high Beta diversity shows that the forests of Suriname and the Guiana Shield are not nearly as homogenous as is often assumed, and consequently protecting this varied biodiversity requires conserving many different forest areas.

The high abundance of several large-bodied dung beetle species in the region is indicative of the intact wilderness that remains. These species support healthy ecosystems through seed dispersal, parasite regulation and other processes. Maintaining continuous primary forest and regulating hunting (such as through hunting-restricted reserves) in the region will be essential for conserving dung beetle communities and the ecological processes they sustain. These results indicate that the intact headwater region of the Upper Palumeu watershed merits the highest conservation priority.

INTRODUCTION

Dung beetles (Coleoptera: Scarabaeidae: Scarabaeinae) are an ecologically important group of insects. By burying dung as a food and nesting resource, dung beetles contribute to several ecological processes and ecosystem services that include: reduction of parasite infections of mammals, including people; secondary dispersal of seeds and increased plant recruitment; recycling of nutrients into the soil; and decomposition of dung as well as carrion, fruit and fungus (Nichols et al. 2008). Dung beetles are among the most cost-effective of all animal taxa for assessing and monitoring biodiversity (Gardner et al. 2008a), and consequently are frequently used as a model group for understanding general biodiversity trends (Spector 2006). Dung beetles show high habitat specificity and respond rapidly to environmental change. Since dung beetles primarily depend on dung from large mammals, they are excellent indicators of mammal biomass and hunting intensity. Dung beetle community structure and abundance can be rapidly measured using standardized transects of baited traps, facilitating quantitative comparisons among sites and studies (Larsen and Forsyth 2005).

METHODS

I sampled dung beetles at both primary camp sites (Upper Palumeu (Juuru camp) and Kasikasima), as well as on top of the Grensgebergte mountains, using standardized pitfall trap transects (see Executive Summary for site details). Ten traps baited with human dung were placed 150 m apart along a linear transect at each site (see Larsen and Forsyth 2005 for more details), except the mountaintop where only six traps could be placed due to space constraints. Traps consisted of 16 oz plastic cups buried in the ground and filled with water with a small amount of liquid detergent. A bait wrapped in nylon tulle was suspended above the cup from a stick and covered with a large leaf. At each site, except the mountaintop (two days), traps were collected every 24 hours for four days, and were re-baited after two days. I set three flight intercept traps at each site to passively collect dung beetle species that are not attracted to dung. I also placed additional pitfall traps whenever possible with other types of baits that included rotting fungus, carrion, dead millipedes, and injured millipedes. All traps were collected daily. I opportunistically collected dung beetles that I encountered in the forest, usually perching on leaves during both day and night. Beetle specimens are deposited at the National Museum of Natural History at the Smithsonian Institution in Washington, DC, USA and will also be deposited at the National Zoological Collection of Suriname in Paramaribo.

To estimate total species richness at each site and assess sampling completeness, I compared the observed number of species with the expected number of species on the basis of randomized species accumulation curves computed in EstimateS (version 7, R. K. Colwell, http://purl.oclc.org/estimates) (Colwell and Coddington 1994). I used an abundance-based coverage estimator (ACE) because it accounts for species abundance as well as incidence, providing more detailed estimates. I also used EstimateS to calculate similarity among sites, using the Morisita-Horn similarity index which incorporates species abundance as well as incidence.

RESULTS AND DISCUSSION

I collected 107 species and 4,483 individuals at the study sites (Table 5.1, Appendix 5.1). I encountered ten species that appear to be new to science (including *Canthidium*, *Dichotomius* and *Uroxys*), and I estimate an additional 10–20 species are undescribed pending further taxonomic revisions. Species richness and abundance was highest around Juuru camp, intermediate at Kasikasima, and lowest on the Grensgebergte mountaintop. Species accumulation curves for dung-baited pitfall traps (based on

abundance-based coverage estimator) indicated that I sampled an estimated 91% of all coprophagous species occurring in the area (Table 5.1). However, sampling completeness was lowest on the Grensgebergte Mountain where sample size was limited by logistical constraints, and I sampled only 68% of the dung-feeding species likely to occur at the site (Table 5.1, Fig. 5.1).

In addition to coprophagous species, I captured 31 dung beetle species that appear not to feed on dung at all, including those attracted only to carrion or to dead invertebrates, fruit, or fungus (Appendix 5.2). Additional species were sampled only in flight intercept traps, and many of these species are poorly represented in collections because they are difficult to sample and in some cases, their diet is unknown (Appendix 5.2). Some of these species show unusual specializations, such as millipede predation or colonization of leaf-cutter ant nests (see interesting species discussion below).

Upper Palumeu, the site with the highest dung beetle abundance and biomass, was also the most isolated from human communities and hunting pressures, suggesting that hunting has had a mild to moderate impact on dung beetle populations around Kasikasima and elsewhere.

The study sites contained a mixture of widely distributed Amazonian species, species restricted to the northern Amazon, Guiana Shield endemics, and some species with even

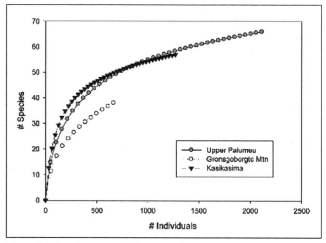

Figure 5.1. Species accumulation curves for each site based on dung-baited pitfall traps

Table 5.1. Diversity and abundance of dung beetles in Grensgebergte and Kasikasima regions

	All sites	Upper Palumeu	Grensgebergte	Kasikasima
Species richness (all samples)	107	93	40	74
Species richness (dung traps)	76	66	38	57
Estimated richness (ACE) (dung traps)	83	83	56	66
% Sampling completeness (dung traps)	0.91	0.80	0.68	0.87
Shannon diversity (H) (dung traps)	3.14	3.05	1.89	3.08
Abundance/trap (all samples)	33.5	37.3	48.6	24.9
Abundance/trap (dung traps)	44.0	52.8	55.3	31.8

more restricted range size. Several species represented new records for Suriname. Species composition varied strongly among sites within this survey, and also among sites sampled on previous RAP surveys in Suriname (Table 5.2). Despite its low overall species richness, the Grensgebergte Mountains appear to support a dung beetle community that is very distinct from the surrounding lowlands. Interestingly, species composition of Grensgebergte was quite similar to the Nassau and Lely Plateaus, and each of these three high elevation habitats are characterized by species that appear to be specialized to these habitats (Table 5.2; Appendix 5.1; Larsen 2007). Surprisingly, species composition of the lowland forests of Upper Palumeu differed fairly strongly from structurally similar lowland forests in nearby Southwestern Suriname around Kwamalasamutu(Table 5.2).

Dung beetle species richness is very high in the Grensgebergte region relative to other areas in northeastern South America and the Guianas, and even higher than the previous record for dung beetle richness in the Guianas in the pristine forests surrounding Kwamalasamutu (Table 5.3; Larsen 2011). Similar RAP surveys at Lely and Nassau in Suriname yielded less than a third of the species richness found around Grensgebergte. Other studies from Venezuela, French Guiana, and Brazil also showed lower species richness in lowland primary forest with comparable sampling effort.

INTERESTING SPECIES

Several species appear to be new to science, including *Canthidium* cf. *minimum* (see page 22), *Canthidium* sp. 25 (miscellum grp.), *Canthon* sp. 2, *Dichotomius* sp. fissus grp. 1 (and probably the other fissus grp. species as well), and *Uroxys* sp. 1. *Canthidium* cf. *minimum* and *Canthidium* sp. 25 (miscellum grp.) also appear to represent new genera to science.

Table 5.2. Dung beetle community similarity among sites within this survey and compared with other RAP surveys throughout Suriname

1st	2nd	S 1st	S 2nd	Shared Species	Morisita-Horn
Upper Palumeu	Grensgebergte	93	40	37	0.28
Upper Palumeu	Kasikasima	93	74	62	0.81
Grensgebergte	Kasikasima	40	74	33	0.46
Upper Palumeu	Kwamala region	93	94	71	0.67
Upper Palumeu	Nassau	93	27	24	0.39
Upper Palumeu	Lely	93	38	28	0.27
Grensgebergte	Kwamala region	40	94	29	0.08
Grensgebergte	Nassau	40	27	16	0.87
Grensgebergte	Lely	40	38	20	0.95
Kasikasima	Kwamala region	74	94	57	0.40
Kasikasima	Nassau	74	27	25	0.48
Kasikasima	Lely	74	38	28	0.42
Kwamala region	Nassau	94	27	21	0.36
Kwamala region	Lely	94	38	28	0.13
Nassau	Lely	27	38	22	0.91

S: Species Richness; Morisita-Horn represents community similarity

Table 5.3. Comparison of dung beetle species richness in primary lowland forests in northeastern South America

	Grensgebergte region	Kwamala region[1]	Nassau, Suriname[2]	Lely, Suriname[2]	Guri, Venezuela[3]	Nouragues, F. Guiana[4,5]	Kaw Mtn, F. Guiana[5]	Jari, Amapa, Brazil[6]	Marajoara, Para, Brazil[7]
S (all samples)	107	94	27	38	41				
S (dung traps)	76	68	24	33	24 (32)	42 (78)	33 (47)	41-51 (72)	47

First number indicates species richness observed with comparable sampling effort to this RAP survey. Number in parentheses indicates species richness observed with more extensive long-term sampling effort, or across a broader landscape. [1]Larsen 2011; [2]Larsen 2007; [3]Larsen et al. 2008, Larsen unpub. data; [4]Feer 2000; [5]Price & Feer 2012; [6]Gardner et al. 2008b; [7]Scheffler 2005

Several large-bodied dung beetle species, such as *Coprophanaeus lancifer* (the largest Neotropical dung beetle species) and *Dichotomius boreus*, were commonly sampled at both lowland sites. These species move long distances and require large, continuous areas of forest to persist. Their presence at the sites is indicative of the intact, contiguous landscape in the region. These large dung beetle species are also the most ecologically important for burying seeds and controlling parasites.

Deltochilum valgum is a highly specialized predator of millipedes, and adults decapitate and feed on millipedes that are much larger than themselves. This unusual behavior was only discovered and described recently (Larsen et al. 2009). *Canthidium* cf. *chrysis* is a member of the escalerei species group which commonly feed on dead invertebrates. It was captured mostly with dead millipedes, but occasionally with carrion, and may be specialized to feed on millipedes. *Anomiopus globosus* and *Anomiopus parallelus* are two unusual species with unknown natural history, although their morphology suggests an association with ant nests.

Several species represent new records for Suriname, including *Ateuchus oblongus*, *Eurysternus howdeni*, and *Feeridium woodruffi*. The presence of *E. howdeni* is very surprising, and represents a massive extension of its known range. *F. woodruffi* was only described in 2008, and is an unusual and extremely rare species known only from a few specimens (Vaz-de-Mello 2008). It is the only species in its genus.

CONSERVATION RECOMMENDATIONS

The Grensgebergte area supports vast tracts of intact primary forest, which is important for many dung beetle species. Consequently, I found extremely high species richness of dung beetles in the area, and the highest known record for dung beetle diversity in the Guiana Shield (107 species). To put this diversity into perspective, during a RAP survey at the Nassau and Lely plateaus in Suriname, I sampled only 24 species and 33 species at each site respectively, and in the Kwamala region, I sampled a total of 94 species across three distinct sites (Table 5.3). As dung beetles are highly sensitive to forest loss and degradation, preventing mining operations and other drivers of deforestation from entering the area will be important for maintaining the high biodiversity of the region.

In addition to high overall species richness, I found high Beta diversity at the sites across very small spatial scales. Consequently, it is important to protect the diversity of soils and habitats that occur in Southeastern Suriname even at small spatial scales. Plans for protected areas or reserves should incorporate this small-scale spatial heterogeneity, as well as taking into consideration that Southeastern Suriname is quite distinct from southwestern and other parts of Suriname.

High dung beetle diversity and biomass at Upper Palumeu, the most isolated site, may be explained by lower hunting pressures. The abundance and biomass of dung beetles in the Grensgebergte area overall was relatively high, and was higher than observed at Nassau, Lely, Kwamala, and other parts of Suriname. This suggests that in addition to the pristine state of the forest, populations of large birds and mammals are relatively stable. However, it is important to regulate hunting intensity, and maintain healthy populations of large mammals such as spider monkeys, howler monkeys and peccaries, as these are among the most important species for dung beetles. Reduced hunting on large, key species would help to stabilize ecosystem dynamics not just for dung beetles, but for seed dispersal and other ecological processes as well. The establishment of hunting-restricted areas such as the tourist area at Kasikasima is an excellent way to maintain sustainable populations of large mammals.

LITERATURE CITED

Colwell, R. K. and J. A. Coddington. 1994. Estimating terrestrial biodiversity through extrapolation. Philosophical Transactions of the Royal Society of London B Biological Sciences 345:101–118.

Feer, F. 2000. Dung and carrion beetles of the rain forest of French Guiana: composition and structure of the guild. Annales De La Societe Entomologique De France 36:29–43.

Gardner, T. A., J. Barlow, I. S. Araujo, T. C. Avila-Pires, A. B. Bonaldo, J. E. Costa, M. C. Esposito, L. V. Ferreira, J. Hawes, M. I. M. Hernandez, M. S. Hoogmoed, R. N. Leite, N. F. Lo-Man-Hung, J. R. Malcolm, M. B. Martins, L. A. M. Mestre, R. Miranda-Santos, W. L. Overal, L. Parry, S. L. Peters, M. A. Ribeiro, M. N. F. da Silva, C. D. S. Motta, and C. A. Peres. 2008a. The cost-effectiveness of biodiversity surveys in tropical forests. Ecology Letters 11:139–150.

Gardner, T. A., M. I. M. Hernandez, J. Barlow, and C. A. Peres. 2008b. Understanding the biodiversity consequences of habitat change: the value of secondary and plantation forests for neotropical dung beetles. Journal of Applied Ecology 45:883–893.

Larsen, T. H. 2007. Dung beetles of the Lely and Nassau plateaus, Eastern Suriname. Pages 99–101 *in* L. E. Alonso and J. H. Mol, editors. A rapid biological assessment of the Lely and Nassau plateaus, Suriname (with additional information on the Brownsberg Plateau). Conservation International, Arlington, VA, USA.

Larsen, T. H. and A. Forsyth. 2005. Trap Spacing and Transect Design for Dung Beetle Biodiversity Studies. Biotropica 37:322–325.

Larsen, T. H., A. Lopera, and A. Forsyth. 2008. Understanding Trait-Dependent Community Disassembly: Dung Beetles, Density Functions, and Forest Fragmentation. Conservation Biology 22:1288–1298.

Larsen, T. H., A. Lopera, A. Forsyth, and F. Genier. 2009. From coprophagy to predation: a dung beetle that kills millipedes. Biology Letters 5:152–155.

Larsen, T. H. 2011. Dung beetles of the Kwamalasamutu region, Suriname (Coleoptera: Scarabaeidae: Scarabaeinae). A Rapid Biological Assessment of the Kwamalasamutu region, Southwestern Suriname. B. J. O'Shea, L. E. Alonso and T. H. Larsen. Arlington, VA, Conservation International. 63: 91–103.

Nichols, E., S. Spector, J. Louzada, T. Larsen, S. Amequita, and M. E. Favila. 2008. Ecological functions and ecosystem services provided by Scarabaeinae dung beetles. Biological Conservation 141:1461–1474.

Price, D. L. and F. Feer. 2012. "Are there pitfalls to pitfalls? Dung beetle sampling in French Guiana." Organisms Diversity & Evolution 12(3): 325–331.

Scheffler, P. Y. 2005. Dung beetle (Coleoptera : Scarabaeidae) diversity and community structure across three disturbance regimes in eastern Amazonia. Journal of Tropical Ecology 21:9–19.

Spector, S. 2006. Scarabaeine dung beetles (Coleoptera : Scarabaeidae : Scarabaeinae): An invertebrate focal taxon for biodiversity research and conservation. Coleopterists Bulletin 60:71–83.

Vaz-De-Mello, F. Z. 2008. Synopsis of the new subtribe Scatimina (Coleoptera: Scarabaeidae: Scarabaeinae: Ateuchini), with descriptions of twelve new genera and review of Genieridium, new genus. Zootaxa(1955): 1–75.

Appendix 5.1. Dung beetle species abundance (number individuals collected), including taxonomic notes and amended species list from RAP #43 & 63

Site	Upper Palumeu (Juuru)	Grensgebergte rock	Kasikasima	Kwamala region (3 sites)	Nassau	Lely	Old species name (from RAP #63 Kwamala)
# Species	**93**	**40**	**74**	**94**	**27**	**38**	
Total abundance	**2460**	**681**	**1342**	**4554**	**204**	**906**	
# Trap samples	**66**	**14**	**54**	**193**	**51**	**53**	
Agamopus castaneus Balthasar	1		12	31			
Anomiopus globosus Canhedo	1			2			
Anomiopus parallelus Harold[1]	1			4		1	
Ateuchus cereus Harold	1		1	2			
Ateuchus cf. *sulcicollis* Harold	2		15	5			
Ateuchus murrayi Harold	132	7	142	76	1	1	
Ateuchus oblongus Harold	25						
Ateuchus pygidialis Harold	1			4			
Ateuchus simplex LePeletier & Serville	291		94	285	1	13	
Ateuchus substriatus Harold	12	1	17	57			
Ateuchus sp. 3	1			5			
Ateuchus sp. 4	1			1			
Ateuchus sp. 5	7	2	1	26			
Ateuchus sp. 6 (aff. *murrayi*)			1	3			
Ateuchus sp. 7 (aff. *aeneomicans*)			1	4			
Ateuchus sp. 8 (carbonarius grp.)[2]	2	1	14	28			*Ateuchus* cf. *obscurus* Harold
Canthidium cf. *chrysis* Fabricius			1	22			
Canthidium cf. *kirschi* Harold	13		3	19	1		
Canthidium cf. *minimum* Harold[3]	9			2			
Canthidium cf. *onitoides* Perty	1			1			
Canthidium depressum Boucomont[4]		6			2	20	*Canthidium guyanense* Boucomont
Canthidium deyrollei Harold	14		9	116			
Canthidium dohrni Harold	4		6	7			
Canthidium funebre Balthasar	1						
Canthidium gerstaeckeri Harold	7		3	38	6		
Canthidium gracilipes Harold	6		2	16			
Canthidium latipleurum Preudhomme de Borre	2		3				
Canthidium sp. 5 (aff. *funebre*)	2			6			
Canthidium sp. 6	1	6	15	35		4	
Canthidium sp. 8 (aff. *quadridens*)	3			10			
Canthidium sp. 9	15	6	1	6			
Canthidium sp. 10	47	2	15	2			
Canthidium sp. 12 (aff. *latum*)	3		3	13			
Canthidium sp. 13	1			3			
Canthidium sp. 15	2			2			

table continued on next page

Appendix 5.1. continued

Site	Upper Palumeu (Juuru)	Grensgebergte rock	Kasikasima	Kwamala region (3 sites)	Nassau	Lely	Old species name (from RAP #63 Kwamala)
Canthidium sp. 18 (aff. *bicolor*)	1			25			
Canthidium sp. 19 (aff. *kirschi*)	1			1			
Canthidium sp. 20 (aff. *chrysis*)	7		3	18			
Canthidium sp. 21 (aff. *persplendens*)	3		1				
Canthidium sp. 22 (aff. *chrysis*)			1				
Canthidium sp. 23	1						
Canthidium sp. 24			2				
Canthidium sp. 25 (miscellum grp.)			4				
Canthidium sp. 26 (aff sp. 10)	11	1	2				
Canthidium sp. 27		1					
Canthon bicolor Castelnau	141	1	27	72	2	46	
Canthon quadriguttatus Olivier	9	18	6	1	1	7	
Canthon sordidus Harold	5		3	33	9	19	
Canthon subhyalinus Harold	1						
Canthon triangularis Drury			5	489	13	14	
Canthon vulcanoae Pereira & Martinez[5]	25		5	1			*Canthon semiopacus* Harold
Canthon sp. 2	1			9			
Canthonella silphoides Harold			1	1			
Coprophanaeus jasius Olivier	6	3	3	5			
Coprophanaeus lancifer Linnaeus	16	2	23	4		1	
Deltochilum carinatum Westwood	12		1	4	2	2	
Deltochilum guyanense Boucomont	24	4	5	6	8		
Deltochilum icarus Olivier	3		3	11	1	3	
Deltochilum orbiculare Lansberge	4	1	1		3	1	
Deltochilum septemstriatum Paulian	14		3	14	4		
Deltochilum valgum Burmeister	1			4			
Deltochilum sp. 1	4	2					
Deltochilum sp. 2	2		3				
Deltorhinum guyanensis Genier	1			2			
Dichotomius apicalis Luederwaldt	11	40	12		1		*Dichotomius* sp. 1
Dichotomius boreus Olivier	191	16	118	219	4	7	
Dichotomius cf. *lucasi* Harold	98	11	28	360			
Dichotomius cf. *horridus* Felsche			1	1		·	*Dichotomius* sp. 5 (calcaratus grp)
Dichotomius melzeri Luederwaldt	1		1				
Dichotomius robustus Luederwaldt	6	1	1	3			
Dichotomius subaeneus Castelnau	78	7		1			
Dichotomius worontzowi Pereira[6]	3	1	1	5			*Dichotomius* sp. 4
Dichotomius sp. 3 (batesi-inachus grp)	5		4	2			
Dichotomius sp. fissus grp. 1	2		3				

table continued on next page

Appendix 5.1. continued

Site	Upper Palumeu (Juuru)	Grensgebergte rock	Kasikasima	Kwamala region (3 sites)	Nassau	Lely	Old species name (from RAP #63 Kwamala)
Dichotomius sp. fissus grp. 2	1						
Dichotomius sp. fissus grp. 3	2						
Dichotomius sp. fissus grp. 4			1				
Dichotomius sp. fissus grp. 5	1						
Eurysternus atrosericus Genier			7	86			
Eurysternus balachowskyi Halffter & Halffter	3		7	3	1		
Eurysternus cambeforti Genier		1	11	8		1	
Eurysternus caribaeus Herbst	78	1	48	296	5	16	
Eurysternus cf. *cayennensis* Castelnau[7]	4	47	2				
Eurysternus cyclops Genier	13		17	1	4	17	
Eurysternus foedus Guerin-Meneville	7	2	20	17			
Eurysternus hamaticollis Balthasar	20			2			
Eurysternus howdeni Genier	1	1					
Eurysternus hypocrita Balthasar	108	70	25		1	1	
Eurysternus ventricosus Gill	21	1	21	3		1	
Feeridium woodruffi Vaz-de-Mello	1		2				
Hansreia affinis Fabricius	189	368	199	97	88	569	
Onthophagus cf. *xanthomerus* Bates	20		23	22			
Onthophagus haematopus Harold	233		6	929	34	52	
Onthophagus rubrescens Blanchard	259	5	177	288	1	11	
Oxysternon durantoni Arnaud	16	2	4	56		24	
Oxysternon festivum Linnaeus	1		16	39			
Oxysternon spiniferum Castelnau	1			3			
Phanaeus bispinus Bates	1			1			
Phanaeus cambeforti Arnaud	3	1		41			
Phanaeus chalcomelas Perty	45	2	20	336	2	7	
Scybalocanthon pygidialis Schmidt	44	8	42		1	10	
Sulcophanaeus faunus Fabricius	10	2	3	1			
Sylvicanthon cf. *securus* Schmidt	7	27		5		4	
Trichillum pauliani Balthasar	10		1	23			
Uroxys pygmaeus Harold	16		29	45	4	1	
Uroxys sp. 1	8	2	2		4	2	
Uroxys sp. 3	40	1	29	81	6	29	

Additional species sampled on other RAP surveys in Suriname

Anomiopus andrei Canhedo				1			
Anomiopus lacordairei Waterhouse				3	1		
Canthidium cf. *gigas* Balthasar				1			

table continued on next page

Appendix 5.1. continued

Site	Upper Palumeu (Juuru)	Grensgebergte rock	Kasikasima	Kwamala region (3 sites)	Nassau	Lely	Old species name (from RAP #63 Kwamala)
Canthidium guyanense Boucomont[4]				7			*Canthidium* sp. 11 (aff. *guyanense*)
Canthidium splendidum Preudhomme de Borre				11			
Canthidium sp. 3						3	
Canthidium sp. 7 (aff. *histrio*)				2			
Canthidium sp. 14 (centrale grp)				2			
Canthidium sp. 16				2			
Canthidium sp. 17				1			
Canthon mutabilis Lucas						3	
Canthon sp. 1				6			
Coprophanaeus dardanus MacLeay						3	
Coprophanaeus parvulus Olsoufieff				2		1	
Dendropaemon sp. 1				3			
Dichotomius mamillatus Felsche				4		1	
Dichotomius sp. 2				3			
Eurysternus vastiorum Martinez						2	
Oxysternon silenus Castelnau						2	
Uroxys gorgon Arrow				1			

[1]As with individuals collected on the RAP survey around Kwamalasamutu, the individual of *A. parallelus* collected here is much larger than those revised by Canhedo (2006), and are also larger and differ in pronotal patterning from the individual from the Lely RAP survey. These specimens may represent an undescribed species.

[2]The name used in the previous RAP survey, *Ateuchus* cf. *obscurus* is apparently incorrect, as *Canthidium obscurum* is a valid species within *Canthidium*. This species belongs to the carbonarius group and may be undescribed.

[3]As described in RAP 63 from Kwamala, *Canthidium* cf. *minimum* is a curious species. It closely resembles *Canthidium minimum*, although the hind tibia is slightly less curved in the specimens from this survey, and all specimens from this survey are much smaller in size. *Canthidium minimum* shares characters with two genera, *Canthidium* and *Sinapisoma*, and might need to be transferred to *Sinapisoma*. *Sinapisoma* is currently a monospecific genus, and the only known species possesses a more elongate and curved inner margin of the hind tibia (which until recently caused it to be erroneously considered a canthonine roller) than in *C. minimum*. However, the hind tibia of *C. minimum* is more elongate and curved than other *Canthidium* species. Both share other characters, including a narrow mesosternum. I believe the species collected here is likely to be undescribed.

[4]Based on comparison of specimens with original type descriptions, it appears that two species from previous RAP surveys were switched (Vaz-de-Mello, pers. comm.). *Canthidium depressum* (misidentified as *C. guyanense* in previous RAP reports) appears to be completely specialized to granite outcrops and plateaus, while *Canthidium guyanense* (identified previously as *Canthidium* sp. 11) is found in lowland forests of southwestern Suriname. The two species are very closely related, but can be separated based on the second and third elytral striae which are deeply impressed posteriorly in *C. depressum*, while the first three elytral striae are all weakly but equally impressed posteriorly in *C. guyanense*.

[5]All specimens identified as *Canthon semiopacus* on previous RAP surveys and also in various museum collections that are from the Guiana shield are probably misidentified, and represent *Canthon vulcanoae* (Vaz-de-Mello, pers. comm.). *C. semiopacus* is a closely related Amazonian species, and can be distinguished by its larger size, red elytra, genitalia, and other characters

[6]The species collected here represents the true *Dichotomius worontzowi*, and is distinct from a common and widespread Amazonian species often misidentified as *D. worontzowi* (Vaz-de-Mello, pers. comm.). *D. worontzowi* has four head tubercles rather than two

[7]The population of *Eurysternus cayennensis* from this survey, which was primarily specialized to the Grensgebergte mountaintop, differs quite strongly from the widespread *E. cayennensis* found farther south and west throughout the Amazon. In addition to a distinct ventral color pattern, the basal tooth and setae of the male hind femur are very distinct. Further comparisons are needed to determine whether this represents an undescribed species.

Appendix 5.2. Diet preference/capture method for dung beetles. Data are number of individuals collected. FIT: flight intercept trap (unbaited).

	Dung	FIT	Dead insects	Dead millipede	Fungus	Fruit
# Species	76	71	15	3	2	4
Total abundance	4045	312	108	3	5	10
# Trap samples	92	29	7	3	1	1
Agamopus castaneus	13					
Anomiopus globosus		1				
Anomiopus parallelus		1				
Ateuchus cereus	1	1				
Ateuchus cf. *sulcicollis*	14	3				
Ateuchus murrayi	266	15				
Ateuchus oblongus		24				1
Ateuchus pygidialis		1				
Ateuchus simplex	338	17	25			5
Ateuchus substriatus	29	1				
Ateuchus sp. 3	1					
Ateuchus sp. 4		1				
Ateuchus sp. 5	4	6				
Ateuchus sp. 6 (aff. *murrayi*)		1				
Ateuchus sp. 7 (aff. *aeneomicans*)	1					
Ateuchus sp. 8 (carbonarius grp.)	17					
Canthidium cf. *chrysis*		1				
Canthidium cf. *kirschi*	2	14				
Canthidium cf. *minimum*		9				
Canthidium cf. *onitoides*	1					
Canthidium depressum	6					
Canthidium deyrollei	22	1				
Canthidium dohrni	9	1				
Canthidium funebre	1					
Canthidium gerstaeckeri	8	2				
Canthidium gracilipes		8				
Canthidium latipleurum		5				
Canthidium sp. 5 (aff. *funebre*)	2					
Canthidium sp. 6	21	1				
Canthidium sp. 8 (aff. *quadridens*)		3				
Canthidium sp. 9	17	5				
Canthidium sp. 10	57	7				
Canthidium sp. 12 (aff. *latum*)	1	5				
Canthidium sp. 13	1					
Canthidium sp. 15		2				
Canthidium sp. 18 (aff. *bicolor*)		1				
Canthidium sp. 19 (aff. *kirschi*)		1				
Canthidium sp. 20 (aff. *chrysis*)		6	3	1		
Canthidium sp. 21 (aff. *persplendens*)		4				

table continued on next page

Appendix 5.2. continued

	Dung	FIT	Dead insects	Dead millipede	Fungus	Fruit
Canthidium sp. 22 (aff. *chrysis*)		1				
Canthidium sp. 23		1				
Canthidium sp. 24		2				
Canthidium sp. 25 (miscellum grp.)	4					
Canthidium sp. 26 (aff sp. 10)	14					
Canthidium sp. 27	1					
Canthon bicolor	169					
Canthon quadriguttatus	32	1				
Canthon sordidus	7	1				
Canthon subhyalinus	1					
Canthon triangularis	5					
Canthon vulcanoae	28	2				
Canthon sp. 2		1				
Canthonella silphoides		1				
Coprophanaeus jasius		6	6			
Coprophanaeus lancifer	27		14			
Deltochilum carinatum	3	3	7			
Deltochilum guyanense	20	1	11	1		
Deltochilum icarus	4		2			
Deltochilum orbiculare	6					
Deltochilum septemstriatum	5	1	11			
Deltochilum valgum		1				
Deltochilum sp. 1	5	1				
Deltochilum sp. 2		4	1			
Deltorhinum guyanensis		1				
Dichotomius apicalis	63					
Dichotomius boreus	325					
Dichotomius cf. *lucasi*	78	52	1	1	4	1
Dichotomius cf. *horridus*	1					
Dichotomius melzeri	2					
Dichotomius robustus	7	1				
Dichotomius subaeneus	70	12				3
Dichotomius worontzowi	5					
Dichotomius sp. 3	7	2				
Dichotomius sp. fissus grp. 1		5				
Dichotomius sp. fissus grp. 2		1				
Dichotomius sp. fissus grp. 3		2				
Dichotomius sp. fissus grp. 4		1				
Dichotomius sp. fissus grp. 5		1				
Eurysternus atrosericus	7					
Eurysternus balachowskyi	10					

table continued on next page

Appendix 5.2. continued

	Dung	FIT	Dead insects	Dead millipede	Fungus	Fruit
Eurysternus cambeforti	12					
Eurysternus caribaeus	124	2	1			
Eurysternus cf. *cayennensis*	43	1	9			
Eurysternus cyclops	30					
Eurysternus foedus	28	1				
Eurysternus hamaticollis	20					
Eurysternus howdeni	2					
Eurysternus hypocrita	201	2				
Eurysternus ventricosus	41	2				
Feeridium woodruffi	3					
Hansreia affinis	755		1			
Onthophagus cf. *xanthomerus*	27	9	7			
Onthophagus haematopus	238	1				
Onthophagus rubrescens	434	7				
Oxysternon durantoni	15	7				
Oxysternon festivum	13	4				
Oxysternon spiniferum		1				
Phanaeus bispinus	1					
Phanaeus cambeforti	4					
Phanaeus chalcomelas	65	2				
Scybalocanthon pygidialis	69	15	9		1	
Sulcophanaeus faunus	15					
Sylvicanthon cf. *securus*	33	1				
Trichillum pauliani	11					
Uroxys pygmaeus	45					
Uroxys sp. 1	9	3				
Uroxys sp. 3	69	1				

Chapter 6

Katydids (Orthoptera: Tettigonioidea) of the Grensgebergte mountains and Kasikasima region of Southeastern Suriname

Piotr Naskrecki

SUMMARY

Fifty two species of katydids (Orthoptera: Tettigoniidae) were recorded during a rapid biological assessment of lowland and mid-elevation forests of the Grensgebergte mountains of SE Suriname. At least six species are new to science, and 26 species are recorded for the first time from Suriname, bringing the number of species of katydids known from this country up to 128. The current survey confirms that the katydid fauna of Suriname is rich, yet still very poorly known. Although no specific conservation issues have been determined to affect the katydid fauna, habitat loss in Suriname due to logging and mining activities constitute the primary threat to the biota of this country.

INTRODUCTION

Katydids (Tettigonioidea) are a large superfamily of orthopteroid insect, which includes approximately 8,600 species distributed worldwide. Based on the rate of discovery and numbers of recently described species of these insects at least 2,000-3,000 species remain to be named. In some areas, especially in the humid areas of the circumtropical belt of the globe, as many as 75% of species found there remain to be collected and formally described. One of such areas is the Guiana Shield, including Suriname, where katydids have never been systematically studied. Despite the recent increase in the faunistic and taxonomic work on katydids of the Neotropical region, forests of Suriname remain some of the least explored and potentially interesting areas of South America. Approximately 200 species of the Tettigoniidae have been recorded from countries comprising the Guiana shield (e.g., Venezuela, Guyana, Suriname, and French Guiana), but this number most likely represents only a small fraction of the regional species diversity, and at least 300–500 species can be expected to occur there. Up to date, 128 species have been reported from Suriname, which includes 29 species collected for the first time in Suriname during a RAP survey in 2010 (Naskrecki 2012), and the species recorded during the present survey. The remainder of these records is based on material collected in the 19th century, and most of the species from Suriname were described in the monographic works by Brunner von Wattenwyl (1878, 1895), Redtenbacher (1891), and Beier (1960, 1962). More recently Nickle (1984), Kevan (1989), Naskrecki (1997), Emsley and Nickle (2001), and Montealegre and Morris (2003) described additional species from the region.

Many katydid species exhibit strong microhabitat fidelity, low dispersal abilities (Rentz 1993), and high sensitivity to habitat fragmentation (Kindvall and Ahlen 1992) thereby making them good indicators of habitat quality and disturbance. These insects produce species-specific acoustic signals, which can be used for non-invasive, remote assessment of species richness and abundance (Diwakar et al. 2007). These insects also play a major role in many terrestrial ecosystems as herbivores and predators (Rentz 1996). It has been demonstrated that katydids are the principal prey item for several groups of invertebrates and vertebrates in Neotropical forests, including birds, bats (Belwood 1990), and primates (Nickle and Heymann 1996).

The following report presents results of a survey of katydids conducted during March 9-28, 2012 at four lowland rainforest sites in the southeastern region of Suriname.

METHODS AND STUDY SITES

During the survey 3 methods were employed for collecting katydids: collecting at an ultraviolet (UV) light at night, visual searches at night and during the day, and detection of stridulating individuals using an ultrasound detector (Petersson D1000X) at night. Representatives of all encountered species were collected and voucher specimens were preserved in 95% ethanol. Voucher specimens of all collected species will be deposited in the National Zoological Collection of Suriname, Paramaribo, while remaining specimens will be deposited in the collections of the Museum of Comparative Zoology, Harvard University and the Academy of Natural Sciences of Philadelphia (the latter will also become the official repository of the types of any new species encountered during the present survey upon their formal description.)

In addition to physical collection of specimens, wing stridulation of several acoustic species was recorded using a Pettersson D1000X digital ultrasound recorder, which allowed for documentation of sound frequencies up to 250 kHz. Virtually all species encountered were photographed, and these images will be available online in the database of the world's katydids (Otte and Eades 2013).

Simpson's Index of Diversity (D_s) was calculated for each site using the formula:

$$D_s = 1 - \sum_1^i [n_i(n_i-1)]/[N(N-1)]$$

where n_i = number of individuals of species i, and N = number of all collected individuals.

Katydids were surveyed at the following four sites:

(**Site 1**) Palumeu Village (N 3.348456, W 55.439417)— 8–9 and 28–29 March 2012— This site was dominated by highly disturbed grassland and a low secondary forest. All katydids found at this site represented genera typical of open, often anthropogenic and disturbed habitats.

(**Site 2**) Upper Palumeu River (Juuru Camp) (N 2.47700 W 55.62941) — 9–19 March 2012— This densely forested site spanned an elevation gradient of 270-500 m a.sl., with the lower portions subject to seasonal inundation by the Palumeu River (including a flood during the survey).

(**Site 3**) Grensgebergte (Rock) (N 2.46554 W 55.77000)— 17–18 March 2012— A large granite inselberg, partially covered by low forest. This site had an exceptionally low diversity and abundance of katydids.

(**Site 4**) Kasikasima Camp (N 2.97731 W 55.38500)— 20–28 March 2012— A lowland, partially inundated site covered with old-growth forest.

RESULTS

Fifty two species of katydids were recorded during the survey, representing 35 genera and 4 subfamilies. At least 6 of the collected species are new to science, including an unusual sylvan katydid lacking the stridulatory organs in the male, which warrants the creation of a new genus. A number of species belonging to genera in need of taxonomic revisions have been identified to the generic level only. A complete list of recorded species and their relative abundance at each site is presented in Appendix 6.1.

Of the four sampling sites, the Kasikasima Camp had the highest number of species (34) followed by the Juuru Camp (29 species); additionally, 5 species were recorded in the Palumeu Village, and only 2 species were collected on the Grensgebergte. Simpson's Indexes of Diversity (D_s) for Sites 1, 2, and 4 were 0.745, 0.9356, and 0.9433, respectively (D_s was not calculated for Site 3 because of an inadequate sampling effort at the site). These results indicate high species richness, combined with low abundance of individual species at Sites 2 and 3, a situation typical of tropical habitats with low levels of disturbance.

Similarly to the previous RAP survey in the Kwamalasamutu region of southern Suriname in 2010 (Naskrecki 2012), during which 78 species were recorded, the abundance of katydids encountered during this survey appeared low compared to that typically encountered in lowland Neotropical forests. Although no formal structured sampling was conducted, the rate of katydid collection was often lower than 1 individual/hour, and during most nights no individuals were attracted to the UV light. This may be related to the fact that the two main sampling sites (Jururu Camp and Kasikasima Camp) were located in a seasonally inundated forest, thus limiting the number of species associated with the forest floor and the lower layers of the forest's understory. It is likely that most of katydid diversity at these sites is concentrated in the forest canopy, a habitat difficult to sample without either direct access to it or the canopy fogging. The use of the UV light within the confines of the camp, where it was largely hidden by the camp's infrastructure, proved quite ineffective for attracting canopy species. Seasonal inundation does not explain, however, the exceptionally low number of katydids found on Grensgebergte, an elevated formation partially covered with dense primary forest that is never subject to flooding. There the low richness and abundance of these insects might be the result of lower temperatures and higher winds than in the surrounding lowlands.

Conehead katydids (subfamily Conocephalinae)

The Conocephalinae, or the conehead katydids, include a wide range of species found in both open, grassy habitats, and high in the forest canopy. Many species are obligate semenivores (seed feeders), while others are strictly predaceous. A number of species are diurnal, or exhibit both diurnal and nocturnal patterns of activity. Fifteen species of this family were recorded, including at least 3 species new to science; 4 species could only be identified to the generic level because of the poor understanding of species boundaries within their genera, which are in dire need of taxonomic revisions.

Moncheca spp.—Two species of this genus were collected during the survey, *M. bisulca* (Serville) and *M.* sp. n. 1. These brightly colored insects, along with species of the closely related genus *Vestria,* are some of the very few katydids known to employ chemical defenses (methapyrazines), which are effective at repelling bird and mammalian predators (Nickle et al. 1996). The compound are released from a dorsal abdominal gland located between the 8th and 9th tergite; in *M.* sp. n.1 the gland was bright red when everted, forming distinctive "lips", which are likely to play an aposematic function. In *M. bisulca* the gland was not colored, but the evertion of the gland's lip was accompanied with fanning of the wings, which revealed striking, brightly yellow coloration of the dorsal surface of the abdomen and red portions of the ovipositor.

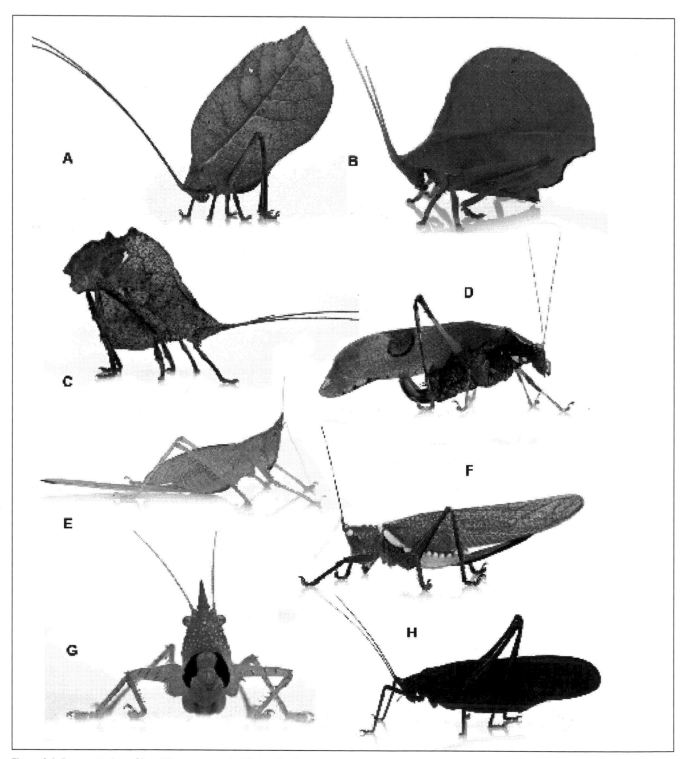

Figure 6.1. Representatives of katydid species recorded during the Grensgebergte Mountains RAP in SE Suriname: (A) *Cycloptera speculata*; (B) *Roxelana crassicornis*; (C) *Typophyllum* sp. 1; (D) *Hetaira smaragdina*; (E) *Copiphora* sp. n.; (F) *Moncheca bisulca*; (G) *Copiphora longicauda*; (H) *Chondrosternum triste.*

Copiphora spp.—Two species of the genus were collected, *C. longicauda* Serville and *C.* sp. 1, and the latter species is new to science. These insects are predators of insects and other invertebrates

Subria sp. 1—This arboreal, predaceous species was collected in the vicinity of both main camps, and is likely new to science.

Artiotonus sp. 1—A single male of this yet undescribed species was collected at the Kasikasima site (see page 24). Interestingly, the genus *Artiotonus* was only described in 2011 (Montealegre et al. 2011) to include 3 species from Colombia and Ecuador. The current record is the easternmost record for this genus.

Leaf katydids (subfamily Phaneropterinae)

The Phaneropterinae, or leaf katydids, represent the largest, most species-rich lineage of katydids, with nearly 2,700 species worldwide, and at least 550 species recorded from South America. All species of this family are obligate herbivores, often restricted to a narrow range of host plants. Probably at least 50–75% of species found in lowland rainforests are restricted to the canopy layer and never descend to the ground (females of many species lay eggs on the surface of leaves or stems, and the entire nymphal development takes place on a single host plant.) For this reason, these insects are difficult to collect, and the only reliable method for their collection is a UV or mercury-vapor lamp, or canopy fogging. Few species can be encountered during a visual or acoustic search in the understory of the forest.

Sixteen species of leaf katydids were recorded during the present survey, virtually all attracted to the UV light at the camps. One species (*Euceraia* sp. 1) is possibly new to science and several species represent records new to Suriname (e.g., *E. abnormalis,* previously known only from Brazilian states Mato Grosso and Minas Gerais.)

Sylvan katydids (subfamily Pseudophyllinae)

Virtually all members of tropical Pseudophyllinae occur only in forested, undisturbed habitats, and thus have a potential as indicators of habitat disturbance. These katydids are mostly herbivorous, although opportunistic carnivory has been observed in some species (e. g., *Panoploscelis*). Many are confined to the upper layers of the forest canopy and never come to lights, and are therefore difficult to collect. Fortunately, many species have loud, distinctive calls, and it is possible to document their presence based on their calls alone, a technique well known to ornithologists and recently applied to monitoring of katydids and other orthopteroid insects (Diwakar et al. 2007). Nineteen species of this family were collected during the present survey.

Gnathoclita vorax (Stoll, 1813)—This spectacular species is a rare example of a katydid with strong sexual dimorphism manifested in strong, allometric growth of the male mandibles. It was found in relative high numbers at both main campsites, especially among the stands of the bamboo *Guadua* sp., the stems of which this species uses as shelter.

G. vorax is known only from southern Guyana and southern Suriname.

cf. *Macrochiton* sp. 1—Numerous individuals of this new genus and species of sylvan katydids were collected at all three camps of the survey. It is a large, macropterous insect, with exceptionally long appendages (see page 24). The unique feature of this insect is the complete absence of the tegminal stridulatory organs in the male, a condition exceptionally rare among the Tettigonioidea. These katydids, being unable to produce wing stridulation, are therefore silent, or employ yet unknown acoustic communication methods. This is the first case in the Neotropics, and only the third known case in the world, of a fully winged katydid devoid of tegminal stridulatory organs. The reasons for loss of this species' ability to produce acoustic signals is a tantalizing mystery, one that should be explored in the future.

Interestingly, this is not the first time this yet unnamed species has been collected: the University of Nebraska insect collection contains several individuals of this species collected in May 1917 in Mana River, French Guiana. Entomologist Lawrence Bruner, who during that time was the curator of the Nebraska insect collection, recognized the uniqueness of this species and designated the specimens as types, but failed to publish their formal description, and thus the species remains undescribed.

?N. gen. and n. sp. (Homalaspidiini)—A single female of a highly unusual member of the poorly known tribe Homalaspidiini was collected at the Juuru Camp (see page 24). The same species (also a single female) was previously collected at Werehpai during the 2010 Suriname RAP. This species bears no resemblance to any known species of the tribe, but until a male specimen is available it will likely remain undescribed.

CONSERVATION RECOMMENDATIONS

The sites visited by the RAP team in Grensgebergte mountains of SE Suriname belong to some of the most pristine, least populated areas in South America, and the results of this survey confirm that the fauna of katydids of southern Suriname includes a large proportion of species remain unknown and unnamed. Despite the relatively low number of species recorded during the current survey, the species turnover between 2010 and 2012 RAP sites was very high: only 26 species (50%) recorded at the Grensgebergte mountains sites had also been found at Kwamalasamutu sites in 2010. This is indicative of biological uniqueness and a possibly high degree of endemism of the Grensgebergte mountains.

More sampling, combined with comprehensive taxonomic and phylogenetic reviews, are badly needed in order to understand its true magnitude. As with most groups of tropical insects, and unlike large mammals and birds, the principal threat to the survival of katydids in Suriname comes not from hunting or trade but rather from habitat alteration and loss, especially from logging and mining.

These insects may be adversely affected by even the most seemingly insignificant changes to the composition of the forest vegetation or humidity, such as those resulting from opening of new roads and the increase of the forest's edge. While species-level conservation recommendations are currently impossible to make, protecting the existing habitats, or at least major, connected fragments of them, is the most effective way of ensuring their survival.

REFERENCES

Beier, M. 1960. Orthoptera Tettigoniidae (Pseudophyllinae II). – In: Mertens, R., Hennig, W. and Wermuth, H. [eds]. Das Tierreich. – 74: 396 pp.; Berlin (Walter de Gruyter and Co.).

Beier, M. 1962. Orthoptera Tettigoniidae (Pseudophyllinae I). – *In*: Mertens, R., Hennig, W. and Wermuth, H. [eds]. Das Tierreich. – 73: 468 pp.; Berlin (Walter de Gruyter and Co.).

Belwood, J.J. 1990. Anti-predator defences and ecology of neotropical forest katydids, especially the Pseudophyllinae. – *In*: Bailey, W.J. and Rentz, D.C.F. [eds]. The Tettigoniidae: biology, systematics and evolution: ix + 395 pp.; Bathurst (Crawford House Press) and Berlin et al. (Springer). Pages 8–26.

Brunner von Wattenwyl, C. 1878. Monographie der Phaneropteriden. 1–401, pls 1–8; Wien (Brockhaus).

Brunner von Wattenwyl, C. 1895. Monographie der Pseudophylliden. IV + 282 pp. [+ X pls issued separately]; Wien (K.K. Zoologisch–Botanische Gesellschaft).

Diwakar, S. M. Jain and R. Balakrishnan. 2007. Psycho-acoustic sampling as a reliable, non-invasive method to monitor orthopteran species diversity in tropical forests. Biodiversity Conservation, 16: 4081–4093.

Emsley, M.G. and Nickle, D.A. 2001. New species of the Neotropical genus *Daedalellus* Uvarov (Orthoptera: Tettigoniidae: Copiphorinae). – Transactions of the American Entomological Society 127: 173–187.

Kevan, D.K.McE. 1989. A new genus and new species of Cocconotini (Grylloptera: Tettigonioidea: Pseudophyllidae: Cyrtophyllinae) from Venezuela and Trinidad, with other records for the tribe. – Bol. Ent. Venez. 5: 1–17.

Kindvall, O. and Ahlen, I. 1992. Geometrical factors and metapopulation dynamics of the bush cricket, *Metrioptera bicolor* Philippi (Orthoptera: Tettigoniidae). – Conservation Biology 6: 520–529.

Naskrecki, P. 1997. A revision of the neotropical genus *Acantheremus* Karny, 1907 (Orthoptera: Tettigoniidae: Copiphorinae). – Transactions of the American Entomological Society 123: 137–161.

Naskrecki, P. 2011. A rapid biological assessment of katydids of the Kwamalasamutu region, Suriname (Insecta: Orthoptera: Tettigoniidae). *In*: O'Shea, B.J., L.E. Alonso, and T.H. Larsen, (eds.). A Rapid Biological Assessment of the Kwamalasamutu region, Southwestern Suriname. RAP Bulletin of Biological Assessment 63. Conservation International, Arlington, VA.

Nickle, D.A. 1984. Revision of the bush katydid genus *Montezumina* (Orthoptera; Tettigoniidae; Phaneropterinae). – Transactions of the American Entomological Society 110: 553–622.

Nickle, D.A., J.L. Castner, S.R. Smedley, A.B. Attygalle, J. Meinwald and T. Eisner. 1996. Glandular Pyrazine Emission by a Tropical Katydid: An Example of Chemical Aposematism? (Orthoptera: Tettigoniidae: Copiphorinae: *Vestria* Stål). Journal of Orthoptera Research, 5: 221–223.

Nickle, D.A. and Heymann E.W. 1996. Predation on Orthoptera and related orders of insects by tamarin monkeys, *Saguinus mystax* and *S. fuscicollis* (Primates: Callitrichidae), in northeastern Peru. – Journal of the Zoological Society 239: 799–819.

Otte, D. and D. Eades. 2012. Orthoptera Species File Online <http://orthoptera.speciesfile.org>

Redtenbacher. 1891. Monographie der Conocephaliden. Verh. der Zoologisch-botanischen Gesellsch Wien 41(2): 315–562.

Rentz, D.C.F. 1993. Orthopteroid insects in threatened habitats in Australia. – Pages 125–138 in: Gaston, K.J., New, T.R. and Samways, M.J. [eds]. Perspectives on Insect conservation: 125–138; Andover, Hampshire (Intercept Ltd).

Rentz, D.C.F. 1996. Grasshopper country. The abundant orthopteroid insects of Australia. Orthoptera; grasshoppers, katydids, crickets. Blattodea; cockroaches. Mantodea; mantids. Phasmatodea; stick insects: i–xii, 1–284; Sydney (University of New South Wales Press).

Appendix 6.1. A checklist of katydids (Orthoptera: Tettigoniidae) recorded during the SE Suriname RAP 2012.

Family	Species	Site			
		Palumeu Village	Juuru Camp	Grensgebergte	Kasikasima Camp
		Individuals of species per site			
Conocephalinae 9 gen., 16 spp.	*Agraecia viridipennis* Redtenbacher		1		
	Subria grandis (Walker)				3
	Subria sp. 1		3		5
	Uchuca sp. 1				3
	Uchuca macroptera Montealegre et al.				8
	Conocephalus sp. 1	5			
	Conocephalus sp. 2	1			
	Conocephalus (Xiphidion) cinereus Thungberg				1
	Artiotonus sp. 1				1
	Copiphora longicauda Serv.		5		7
	Copiphora sp. 1		3		4
	Graminofolium castneri Nickle		6		1
	Moncheca bisulca (Serv.)		1	1	18
	Moncheca sp. 1		1		
	Neoconocephalus sp. 3 Suriname		1		1
	Neoconocephalus sp. 4 Suriname	1		1	
Listroscelidinae 1 gen., 1 sp.	*Phlugis* cf. *teres*	1			
Phaneropterinae 11 gen., 16 spp.	*Paraphidnia verrucosa* (Br. Watt.)		1		
	Steirodon dentatum (Stål)		2		
	Steirodon degeeri (Stål)		1		2
	Stilpnochlora undulata Emsley				1
	Anaulacomera sp. 1		6		
	Ceraia sp. 1				2
	cf. *Terpnistroides* sp. 1		1		
	Euceraia sp. 1		1		
	Euceraia abnormalis (Br. Watt.)		1		1
	Hetaira smaragdina Br. Watt.		1		
	Parableta sp. 1 Suriname	1			1
	Phylloptera festae Griffini		2		
	Phylloptera neotenella Otte				1
	Phylloptera sp. 1				3
	Phylloptera sp. 4		1		
	Proviadana sp. 1		1		
Pseudophyllinae 15 gen., 19 spp.	*Schedocentrus* sp. 1				2
	Schedocentrus (S.) vicinus Beier		1		1
	Gnathoclita vorax (Stoll)		6		2
	Gen_Homalaspidini sp. 1		1		
	Chondrosternum triste Beier		17		7

table continued on next page

Appendix 6.1. continued

Family	Species	Site			
		Palumeu Village	Juuru Camp	Grensgebergte	Kasikasima Camp
		Individuals of species per site			
	Leptotettix falconarius (De Geer)				2
	cf. *Macrochiton* n. gen and n. sp.		8		3
	Triencentrus sp. 1		1		
	Rhinischia regimbarti (Griffini)				2
	Rhinischia sp. 1		1		
	Cycloptera speculata (Burm.)		1		
	Pterochroza ocellata (L.)				2
	Roxelana crassicornis (STål)		2		
	Typophyllum sp. 1		4		2
	Typophyllum sp. 2		3		1
	Eumecopterus incisus Beier		4		7
	Teleutias sp. 1				1
	Teleutias sp. 2		5		1
	Teleutias surinamus Beier				2
	Total specimens (species)	12 (5)	93 (30)	2 (2)	98 (34)

Chapter 7

A rapid assessment of the ants of the Grensgebergte and Kasikasima regions of Southeastern Suriname

Leeanne E. Alonso and Jackson A. Helms

SUMMARY

A total of 149 ant species from 35 genera and 10 subfamilies have been identified from the collections made during the 2012 RAP survey of Southeastern Suriname. Additional work is ongoing to process and identify the remaining samples, which will undoubtedly raise the total number of species, possibly to over 200 species. The results indicate a healthy and diverse ant fauna reflective of pristine rainforest. Ants play important roles as predators, scavengers, and seed-dispersers in tropical forests. The ant data from Southeastern Suriname will add to a growing dataset on the ant ant fauna of the Guiana Shield, which is still poorly documented, to help identify areas of high diversity and endemism that are important to conserve within the region. Data on ants and other invertebrates are important since these groups may be able to illustrate differences between habitats within the Guiana Shield that larger animals with wide geographical ranges do not discern.

INTRODUCTION

With over 12,000 described species of ants in the world (see AntWeb), and their social lifestyle consisting of colonies ranging in size from just a few to millions of workers, ants are a dominant force in all terrestrial ecosystems, especially tropical rainforests. They are important members of the ecosystem, with high biomass and population size, and provide key ecological functions such as aerating and turning soil, dispersing plant seeds, consuming dead animals, and controlling pest insects. In addition to their ecological importance, ants have several features that make them especially useful for conservation planning, including: 1) they are dominant members of most terrestrial environments, 2) they are easily sampled in sufficiently high numbers for statistical analysis in short periods of time (Agosti et al. 2000), 3) they are sensitive to environmental change (Kaspari and Majer 2000), and 4) they are indicators of ecosystem health and of the presence of other organisms, due to their numerous symbioses with plants and animals (Alonso 2000).

Ants have been poorly surveyed in Suriname. After the 2005 RAP survey of the Lely and Nassau Mountains (Sosa Calvo 2007) and the 2010 RAP survey in the Kwamalasamutu region of SW Suriname (Alonso 2011), a conservative estimate of about 370 ant species had been recorded in Suriname. However, given the low effort of ant sampling in Suriname and the few localities sampled, there are likely many more ant species in the country.

METHODS

Ants were surveyed at RAP Site 1 (Upper Palumeu River), RAP Site 2 (Grensgebergte), and RAP Site 4 (Kasikasima). Ants were surveyed using hand search-collecting methods and the Winkler method (Agosti et al. 2000). In the search-collecting method, the ants nesting under stones, under or inside decayed wood and those foraging on ground, litter, tree trunk and plants were searched for and collected. This method was employed at all three camps around the camp and in the forest along the principal trails.

The second sifting method used was the Ants of the Leaf Litter (ALL) protocol (Agosti et al. 2000). Along each transect, a 1×1-m quadrat was set up every 10 m (for a total of 10 quadrats per transect). The leaf-litter, rotten twigs, and first layer of soil present in the quadrat were collected into a cloth sifter and shaken for about a minute. Within the sifter was a wire sieve of 1-cm^2 mesh size, which allowed small debris and invertebrates such as ants to fall through the mesh into the bottom of the sifting sack. The sifted leaf litter was then placed in a full-sized Winkler sack, which is a cotton bag into which four small mesh bags containing the leaf litter are placed. Due to their high level of activity, ants run out of the litter and the mesh bag and fall to the bottom of the sack into a collecting cup of 95% ethanol. The Winkler sacks were hung in the field lab for 48 hours.

A total of 10 Winkler Transects were sampled at the following sites:

1. Upper Palumeu (RAP Site 1, approx. 270 m a.s.l.): Two 100 m winkler transects were sampled in the forest along the trail to Brazil, starting after the helipad. Leaf litter was dry and fairly thin. Two 100 m winkler transects were sampled in the forest up the hill behind the camp. 5 days collecting.

2. Grensgebergte (RAP Site 2, approx. 800 m a.s.l.): One 100 m winkler transect was sampled in forest behind the RAP tent. Leaf litter was very wet due to heavy rain the night before and in the morning. One 100 m winkler transect was sampled in the forest at the top of the mountain above the rocky outcrops. Litter was wet and thick. These two transects were combined when hung in the Winkler sacs due to logistical difficulties when RAP Camp 1 flooded. 1 day collecting.

3. Kasikasima (RAP Site 4, approx. 200 m a.s.l.): Two 100 m winkler transects were sampled along the trail by the river (towards the METS camp). Three 100 m winkler transects were sampled in forest along the trail between the RAP camp and the Kasikasima rock, most on the tops of hills on flat plateaus. 6 days collecting.

The ant specimens were preserved in 95% ethanol and sorted to morphospecies. Some specimens were identified to species level by L. Alonso and J. Sosa-Calvo using ant taxonomic literature and the ant collection at the National Museum of Natural History in Washington, D.C.

PRELIMINARY RESULTS

For this report, ants from six of the 10 winkler transects were sorted and identified to morphospecies. These included: 2 transects from RAP Site 1 (Upper Palumeu), 2 transects from RAP Site 2 (Grensgebergte), and 3 transects from RAP Site 4 (Kasikasima). Only a few of the hand collecting samples have been identified so far, with an emphasis on the ants collected from ant-plants.

A total of 149 ant morphospecies representing 35 genera and 10 subfamilies were identified from the sorted collections (see Appendix 7.1). Most of the species were from the Winkler transects since only a few of the hand collections have been sorted so far. Additional work is ongoing to process and identify the remaining samples, which will undoubtedly raise the total number of species, possibly to over 200 species. A total of 72 ant species were recorded from RAP Site 1, 25 species from RAP Site 2, and 92 species from RAP Site 4 (Appendix 7.1).

Genera typical of the region including many large ants that were commonly seen in the forest, including the arboreal species *Daceton armigerum*, *Cephalotes* spp., and *Camponotus* spp., the large-eyed terrestrial *Gigantiops destructor*, and several species of army ants. Many species of tiny leaf

litter dwelling dacetines (*Strumigenys, Octostruma*) were collected at all three sites, indicating good primary forest. Species of the genera *Pheidole, Pachycondyla* and *Odontomachus* were commonly observed and collected. *Pheidole* was the most speciose genus, which is typical for tropical rainforest.

Many ant-plants in the family Melastomataceae (*Tococa* sp.?) were found in the area, housing obligate ant mutualists in domatia at the base of leaves. Most of these plants contained *Pheidole* (possibly cf. *minutula*) or *Crematogaster* sp. 7. The ant-plant *Hirtella duckei* (Chrysobalanaceae) was collected at RAP Site 4 and housed *Allomerus* sp. (see page 26) *Pseudomyrmex* sp. was collected from *Triplaris* near the Kasikasima Camp (Site 4), and *Azteca* sp. was collected from *Cecropia* on top of the rock at RAP Site 2.

One or two species of fire ants, *Solenopsis* spp., were collected during the RAP survey. They may both be *S. geminata*, the tropical fire ant, which is native to the region. A light orange species was found under a log in the forest near RAP Camp 1 and in Palumeu village, and a darker species was common on open rocks on top of the Grensgebergte mountain (Site 2) at about 500 m. Further study is needed to determine the species.

DISCUSSION

Ants were abundant at the Southeastern Suriname RAP sites but did not seem as numerous or conspicuous as they were at the Werehpai RAP site during the Kwamalasamutu RAP survey (Alonso 2011). Much of the area surveyed was in seasonally flooded forest, which may partially explain the perceived lower abundance of ants and thin leaf litter in many areas. The forest between the RAP camp at RAP Site 1 (Upper Palumeu River) and the helipad became completely inundated on August 17. Ants and other organisms in the leaf litter are adapted to this environment and often must move when the floodwaters come. *Atta* sp. (leaf cutting ants) nests were observed only on higher terra firme ground.

Tropical lowland rainforests typically harbor a high diversity of ants. For example, Longino et al. (2002) found over 450 ant species in an area of approximately 1500 ha in Costa Rica, and LaPolla et al. (2007) reported 230 species from eight sites in Guyana. A RAP survey in Papua New Guinea (Lucky et al. 2011) reported 177 ant species from the lowland site (500 m). More studies of ant diversity throughout Suriname are needed to estimate the country's ant diversity, and thereby provide important baseline data for conservation and management of Suriname's biodiversity.

While in-depth analysis cannot be done until all the ant samples have been processed, these preliminary results indicate that Southeastern Suriname has high ant diversity. Data from winkler transects sampled at eight sites in Guyana (LaPolla et al. 2007) reveal comparable levels of ant species richness per transect. The winkler transects done in Guyana included 20 samples each while those from this RAP survey included 10 samples each. Thus two transects from this survey would roughly equal the sampling effort of one transect in Guyana. Table 7.1 shows the number of ant species recorded in each of the Winkler transects in this RAP survey and in Guyana at comparable elevations.

Table 7.1. Number of ant species recorded in per Winkler transects during this RAP survey of Southeastern Suriname (this RAP survey) and from Guyana (LaPolla et al. 2007).

Site	Elevation	Transect #	# samples per transect	# ant species
RAP Site 1, SE Suriname	277 m	1	10	27
RAP Site 1, SE Suriname	277 m	4	10	22
RAP Site 2, SE Suriname	800 m	5	20	16
RAP Site 4, SE Suriname	200 m	6	10	20
RAP Site 4, SE Suriname	200 m	8	10	34
RAP Site 4, SE Suriname	200 m	9	10	38
Kanuku Mountains, Guyana	224 m	1	20	55
Base Camp, Guyana	732 m	1	20	63
Dicymbe Camp, Guyana	717 m	1	20	38

Likewise, further analyses are needed to determine if there are differences between the ant species composition of the three RAP sites sampled in Southeastern Suriname, and also between Southeastern Suriname and other sites within the Guiana Shield. However, the preliminary morphospecies data for the six winkler transects sorted suggest that there may be some differences in species composition between the sites (see Appendix 7.1). Many of the leaf litter ant morphospecies identified so far (collected in Winkler transects) were found at only one of the RAP sampling sites. RAP Sites 1 and 3 were sampled for an equivalent amount of time (5 and 6 days respectively) so further comparisons can be made between the two sites.

Altitudinal differences in ant species richness and composition have been well documented in many tropical regions of the world with higher richness at lower elevations (Johnson and Ward 2002, Lessard et al. 2007). Thus it may be that the Grensgebergte site (RAP Site 2, 800 m) has lower ant richness than the two other sites. However, this difference is more likely due to the almost constant rainy conditions during sampling at that site and to the short duration of sampling there (1 day). The site was overcast with clouds during most of the survey. The winkler transect was collected

during rain and thus the leaf litter was very wet and sticky. Furthermore, the leaf litter sample had to be taken to RAP Site 1 to be hung in the winkler sacks, but RAP Camp 1 flooded at that time and thus the sample was hung two days later, which likely affected the survival of the ants in the sample.

Ants play many critical roles in the functioning of the tropical terrestrial ecosystem, including dispersing seeds, tending mutualistic Homoptera, defending plants, preying on other invertebrates and small vertebrates, and modifying the soil by adding nutrients and aeration (Philpott et al. 2010). Another critical function provided by ants is that of scavenging (see page 27); ants are often the first animals to arrive upon a dead animal and start the decomposition process. Ants are particularly important to plants since they move soil along the soil profile through the formation of their mounds and tunnels, which directly and indirectly affects the energy flow, habitats, and resources for other organisms (Folgarait 1998). Thus ants are important to study and to include in conservation planning.

ANT SPECIES OF NOTE

Several ant species in Southeastern Suriname are common and conspicuous and have interesting life histories and behaviors. These ant species can thus serve to highlight the key roles that ants play in the ecosystem.

Gigantiops destructor—the Jumping Ant—is a large black ant common on the forest floor in the Werehpai area. These ants have extremely large eyes with which to see and avoid their predators and their prey. They move very quickly and actually jump around on the leaf litter, which is unusual for an ant. Despite its name—*destructor*—these ants are timid, so you have to sneak up on them carefully. They do not bite or sting but defend themselves by spraying formic acid from their gaster. These ants forage for small invertebrates in the leaf litter and are often found nesting near *Paraponera clavata* nests, possibly to benefit from the aggressive defense of the larger ants.

Daceton armigerum—the Canopy Ant—is a beautiful golden-colored ant that lives high in the canopy of trees. They have large heads with strong muscles that power their sharp mandibles. Their eyes are under their head so that they can see below them as they walk along branches in the treetops. Another key to their success in the canopy is that their claws are very clingy and can keep a tight hold on branches and tree trunks.

Cephalotes atratus—the Turtle Ant, or Gliding Ant, lives high up in the tree canopy. With its flattened body and large turtle-shaped head, it lives within rotting twigs and branches and blocks the entrance to its nest with its head. Living so high in the canopy, these ants face the threat of falling out of their tree into the terrestrial territories of other, more ferocious ants. Thus they have evolved a way to avoid falling to the forest floor. If they fall from their tree, these ants stretch out their bodies and legs to glide (Yanoviak 2005). They can detect the tree trunk by the relative brightness against the dark greenery and twist in the air to point their abdomen toward their host tree, making a safe landing back home.

Eciton burchellii—the Army Ant—has very large colonies with millions of workers that move through the forest in a swarm raid, capturing everything in their path. These ants do not have a permanent nest but have a "bivouac"—a temporary nest site consisting of a giant ball of ants, usually found under a rotting log or in the hollow of a tree. These ants sting and bite and are very aggressive, even to humans, so one needs to watch where they step around these ants. It is very interesting to watch an army ant swarm since many other creatures can be seen jumping and running to get out of the path of the ants, and some specialized antbirds follow the swarm to catch these invertebrates for their meal. The soldiers of *E. burchellii* have very long mandibles that are used to suture wounds by some indigenous peoples. In addition to their swarms for catching food, these ants are also often seen moving their colony to a new bivouac (which is necessary when they run out of food in an area), carrying their larvae and pupae slung under their bodies.

Odontomachus spp.—Trap-Jaw Ants—are large ants common on the forest floor (see page 27). These ants hold the world record for the fastest reflex in the animal kingdom. They forage by walking around with their mandibles (jaws) wide open. They have small trigger hairs between the mandibles that detect prey items (such as small invertebrates) and trigger the mandibles to snap shut very quickly to capture the prey. These ants often nest in the leaf litter trapped in small palm trees, in the terrestrial leaf litter, or in the soil. They are long, sleek, elegant ants, but have a nasty sting, so care must be taken to avoid touching them.

Paraponera clavata—the Bullet Ant or Congo Ant—is famous for its very powerful and painful sting. It is one of the world's largest ant species and is common in Neotropical lowland rainforests. These ants nest in the ground at the base of trees but forage up in the tree-tops on nectar and invertebrates. While they forage solitarily, they often have a relay of ants for passing large nectar droplets from the treetops to the nest, from one ant to another. These ants are one of the few ant species that make sound to communicate with one another. They can "stridulate" by rubbing their legs along their thorax to make a high-pitched squeaky sound.

Atta sp.—the Leaf-cutting Ant—is well known for its unique and fascinating agricultural lifestyle. *Atta* are fungus-growers—the workers cut pieces of leaves from a wide variety of trees to bring back to their nest where the leaves are chewed up by smaller workers and inserted into a large fungus garden, which the ants tend and cultivate. The ants do not feed on the leaves. Instead, they feed the fruiting bodies of the fungus to their larvae. Their nests are very large with many large underground chambers. It is fun to watch the workers cutting leaves and carrying them over their head back to the colony. *Atta* are parasitized by tiny phorid flies, which lay their eggs on the ants. When a fly larvae hatches, it burrows into an ant's head and develops inside, thereby killing the ant. Small *Atta* workers are often seen hitching a ride on the leaf carried by a larger worker- it is thought that these small ants serve to ward off attacking phorid flies.

Pseudomyrmex spp.—the Tree-dwelling Ants. Many species live in the rotting, hollow twigs and branches up in the trees. They often fall from the trees, landing on the top of tents and even on your shirt, especially after a wind blows through the forest. Some species are specialized, obligate inhabitants of ant-plants, which provide a hollow cavity and sometimes food bodies or nectar for the ants. In exchange, the ants protect the plant by capturing and eating herbivorous insects that may eat the plant. These ants have large eyes and very long, slender bodies (their body form is distinctive) and a painful sting, so it's best to take care when observing them.

CONSERVATION DISCUSSION AND RECOMMENDATIONS

The ant data from this RAP survey are part of an ongoing program of the "Ants of the Guiana Shield" led by Dr. Ted Schultz of the Smithsonian's National Museum of Natural History in Washington, D.C. and collaborators. These data will be combined with the winkler data collected by J. LaPolla and colleagues in Guyana (LaPolla et al. 2007), data from previous RAP surveys, and future surveys in the Guiana Shield to determine hotspots of ant diversity and endemism in the Guiana Shield to guide conservation priorities. The data will also be used to select indicator groups that can be used to more quickly assess the status of the ant community and the ecosystem.

Data on ants and other invertebrates are important since these groups may be able to illustrate differences between habitats within the Guiana Shield that larger animals with wide geographical ranges do not discern. This will aid in identifying important areas of the Guiana Shield for conservation. More data should be collected on ants and other invertebrates for the Guiana Shield in order to determine these patterns.

Like many tropical taxa, many ant species and populations face a range of threats. The most immediate and widespread threat comes from the loss, disturbance, or alteration of habitat. Fragmentation studies have revealed that ant species richness and genetic diversity can be affected even in large forest patches of 40 km² (Brühl et al. 2003, Bickel et al. 2006). Nomadic ant species such as army ants need large expanses of habitat to find enough food to feed their exceptionally large colonies (Gotwald 1995). Likewise, deforestation and forest fragmentation can cause local extinctions of the neotropical swarm-raiding army ant *Eciton burchellii* and other army ants (Boswell et al. 1998, Kumar and O'Donnell 2009).

Global climate change is likely already affecting the distribution of many ant species. For example, Colwell et al. (2008) predict that as many as 80% of the ant species of a lowland rainforest could decline or disappear from the lowlands due to upslope range shifts and lowland extinctions (biotic attrition) resulting from the increased temperature of climate change.

Solenopsis geminata is native to South American rainforests but can become a destructive pest when it spreads into disturbed areas or is introduced to other parts of the world. This species was present in disturbed areas in the villages and could become more widespread in the forest if it moves in along trails. This species often invades areas that have been disturbed so their absence is a good sign of a healthy ant fauna and ecosystem. It is recommended to survey and monitor the presence of this species on the trail to Kasikasima to avoid spread of this species by humans.

Given that ants are highly conspicuous and abundant in Southeastern Suriname they should be a key component of nature walks and eco-tourism visits in the region. Several ant species are large enough to attract the attention and admiration of tourists. These ant species have fascinating life histories and behaviors that give them "personalities" which tourists will find fascinating.

Many ant species require closed canopy forest to maintain the appropriate microclimate they need to survive. These species are found only at pristine sites. Preliminary indications of the ant fauna at all three sites indicate the presence of many forest species among the ant fauna. A full analysis of the ant species, once identified, will reveal whether any ant species are of conservation concern and also how some of the ant species can serve as indicators of the health of the ecosystem.

Southeastern Suriname is one of the last extensive pristine rainforests of the world, containing high and unique biodiversity. This region should be protected from fragmentation by development such as roads and hydropower projects. Likewise, mining and other extractive industries should also be prohibited in the region to avoid impacts on the forests, its species, and the freshwater resources of the region.

REFERENCES

Agosti, D., J.D. Majer, L.E. Alonso, T. R. Schultz (eds.). 2000. Ants: Standard Methods for Measuring and Monitoring Biological Diversity. Smithsonian Institution Press. Washington, D.C. USA.

Alonso, L. E. 2000. Ants as indicators of diversity *In:* Ants, Standard Methods for Measuring and Monitoring Biodiversity, D. Agosti, J. Majer, L. E. Alonso and T. R. Schultz (eds.). Washington, DC: Smithsonian Institution Press.

Alonso, LE. 2011. A preliminary survey of the ants of the Kwamalasamutu region, SW Suriname. *In:* B. O'Shea, L.E. Alonso, and T.H. Larsen (eds.), A rapid biological assessment of the Kwamalasamutu region, Southwestern Suriname. RAP Bulletin of Biological Assessment 63. Conservation International, Arlington, VA, USA.

AntWeb. Accessed June 19, 2013. www.antweb.org.

Bickel, T.O., Brühl, C.A., Gadau, J.R., Hölldobler, B., and Linsenmair, K.E. (2006). Influence of habitat fragmentation on the genetic variability in leaf litter ant populations in tropical rainforests of Sabah, Borneo. Biodiversity and Conservation. 15:157–175.

Boswell, G.P., Britton, N.F. and Franks, N.R. (1998). Habitat fragmentation, percolation theory, and the conservation of a keystone species. Proceedings of the Royal Society of London B. 265:1921–1925.

Brühl, C.A., Eltz, T., and Linsenmair, K.E. (2003). Size does matter—effects of tropical rainforest fragmentation on the leaf litter ant community in Sabah, Malaysia. Biodiversity and Conservation. 12:1371–1389.

Colwell, R., G. Brehm, C.L. Cardelús, A.C. Gilman, and J.T. Longino (2008). Global warming, elevational range shifts, and lowland biotic attrition in the wet tropics. Science. 322:258–261.

Folgarait, P.J. (1998). Ant biodiversity and its relationship to ecosystem functioning: a review. Biodiversity and Conservation. 7:1221–44.

Gotwald, W. (1995). Army Ants: The Biology of Social Predation. Cornell University Press, Ithaca, NY, USA.

Johnson, R.A. and P.S. Ward. 2002. Biogeography and endemism of ants (Hymenoptera: Formicidae) in Baja California, Mexico: a first overview. Journal of Biogeography. 29: 1009–26.

Kaspari, M. and J.D. Majer. 2000. Using ants to monitor environmental change. *In:* Ants, Standard Methods for Measuring and Monitoring Biodiversity, D. Agosti, J. Majer, L. E. Alonso and T. R. Schultz (eds.). Washington, DC: Smithsonian Institution Press. USA.

Kumar, A. and O'Donnell, S. (2009). Elevation and forest clearing effects on foraging differ between surface—and subterranean—foraging army. Journal of Animal Ecology. 78:91–97.

LaPolla, J.S., T. Suman, J. Sosa-Calvo, and T.R. Schultz. 2007. Leaf litter ant diversity in Guyana. Biodiversity and Conservation. 16:491–510.

Lessard, J-P, R. R. Dunn, C.R. Parker, and N.J. Sanders. 2007. Rarity and diversity in forest ant assemblages of Great Smoky Mountains National Park. *Southeastern Naturalist*, Special Issue 1, 215–228.

Longino, J.T., J. Coddington, and R.K. Colwell. 2002. The ant fauna of a tropical rain forest: Estimating species richness three different ways. Ecology 83: 689–702.

Lucky, A., E. Sarnat, and L.E. Alonso. 2011. Ants of the Muller Range, Papua New Guinea. *In:* Richards, S.J. and Gamui, B.G. (eds.) Rapid Biological Assessments of the Nakanai Mountains and the upper Strickland Basin: surveying the biodiversity of Papua New Guinea's sublime karst environments. RAP Bulletin of Biological Assessment 60. Conservation International, Arlington, VA.

Philpott, S.M., I Perfecto, I. Armbrecht, and C.L. Parr. 2010. Ant diversity and function in disturbed and changing habitats. *In:* L. Lach, C.L. Parr, and K.L. Abbott (editors), Ant Ecology, Oxford University Press, New York, USA.

Sosa Calvo, J. 2007. Ants of the leaf litter of two plateaus in Eastern Suriname. *In:* Alonso, L.E. and J.H. Mol (eds). A Rapid Biological Assessment of the Lely and Nassau Plateaus, Suriname (with additional information on the Brownsberg Plateau). RAP Bulletin of Biological Assessment 43. Conservation International, Arlington, VA, USA.

Yanoviak, S.P. and M. Kaspari. 2005. Directed aerial descent in canopy ants. Nature. 433: 624–6.

Appendix 7.1. Ants collected during the 2012 RAP survey of Southeastern Suriname. W=winker leaf litter transect, H=hand collecting

Subfamily	Genus	Species	Upper Palumeu (270 m) 5 days sampling, 2 winkler transects	Grensgebergte (800 m) 1 day sampling, 2 winkler transects*	Kasikasima (200 m) 6 days sampling, 3 winkler transects
Amblyoponinae	Amblyopone	sp. 1			W
Amblyoponinae	Amblyopone	sp. 2			W
Dolichoderinae	Azteca	sp. 1		H	
Dolichoderinae	Azteca	sp. 2	H		
Dolichoderinae	Dolichoderus	sp. 1	H	H	
Dolichoderinae	Dolichoderus	sp. 2	H		
Dolichoderinae	Dolichoderus	sp. 3			W
Ecitoninae	Eciton	sp. 1	H		H
Ecitoninae	Eciton	sp. 2	H		H
Ectatomminae	Ectatomma	ruidum			W
Ectatomminae	Ectatomma	sp. 1	H		
Ectatomminae	Gnamptogenys	cf. regularis			W
Ectatomminae	Gnamptogenys	sp. 1		W	
Formicinae	Acropyga	sp. 1	W		
Formicinae	Camponotus	sp. 1			W
Formicinae	Camponotus	sp. 2	H	H	
Formicinae	Camponotus	sp. 3	H		
Formicinae	Camponotus	sp. 4	H		
Formicinae	Gigantiops	destructor	H		H
Formicinae	Nylanderia	sp. 1		W	W
Formicinae	Nylanderia	sp. 2			W
Formicinae	Nylanderia	sp. 3		W	
Formicinae	Nylanderia	sp. 4		H	
Formicinae	Nylanderia	sp. 5	W		W
Formicinae	Nylanderia	sp. 6	W		
Myrmicinae	Allomerus	sp. 1	H		H
Myrmicinae	Apterostigma	sp. 1			W
Myrmicinae	Apterostigma	sp. 2	W		
Myrmicinae	Apterostigma	sp. 3			W
Myrmicinae	Apterostigma	sp. 4	W		
Myrmicinae	Apterostigma	sp. 5	W		
Myrmicinae	Atta	sp. 1	H		H
Myrmicinae	Carebara	inca			W
Myrmicinae	Carebara	brevipilosa			W
Myrmicinae	Cephalotes	sp. 1	H		
Myrmicinae	Crematogaster	sp. 1			W
Myrmicinae	Crematogaster	sp. 2			W
Myrmicinae	Crematogaster	sp. 3		W	
Myrmicinae	Crematogaster	sp. 4	W		W
Myrmicinae	Crematogaster	sp.5	H		

table continued on next page

Appendix 7.1. continued

Subfamily	Genus	Species	Upper Palumeu (270 m) 5 days sampling, 2 winkler transects	Grensgebergte (800 m) 1 day sampling, 2 winkler transects*	Kasikasima (200 m) 6 days sampling, 3 winkler transects
Myrmicinae	*Crematogaster*	sp. 6	H		
Myrmicinae	*Crematogaster*	sp. 7			H
Myrmicinae	*Crematogaster*	sp.8	W		W
Myrmicinae	*Cyphomyrmex*	sp. 1	W		W
Myrmicinae	*Cyphomyrmex*	sp. 2			W
Myrmicinae	*Cyphomyrmex*	sp. 3			W
Myrmicinae	*Daceton*	*armigerum*	H		H
Myrmicinae	*Megalomyrmex*	sp. 1			W
Myrmicinae	*Megalomyrmex*	sp. 2			W
Myrmicinae	*Myrmicocrypta*	*longinoda*			W
Myrmicinae	*Myrmicocrypta*	*buenzlii*	W		
Myrmicinae	*Ochetomyrmex*	*semipolitus*		W	W
Myrmicinae	*Octostruma*	cf. *balzani*		W	W
Myrmicinae	*Octostruma*	sp. 1			W
Myrmicinae	*Pheidole*	sp. 1			W
Myrmicinae	*Pheidole*	sp. 2			W
Myrmicinae	*Pheidole*	sp. 3			W
Myrmicinae	*Pheidole*	sp. 4			W
Myrmicinae	*Pheidole*	sp. 5			W
Myrmicinae	*Pheidole*	sp. 6	W		W
Myrmicinae	*Pheidole*	sp. 7			W
Myrmicinae	*Pheidole*	sp. 8	W		W
Myrmicinae	*Pheidole*	sp. 9			W
Myrmicinae	*Pheidole*	sp. 10			W
Myrmicinae	*Pheidole*	sp. 11			W
Myrmicinae	*Pheidole*	sp. 12			W
Myrmicinae	*Pheidole*	sp. 13	H		W
Myrmicinae	*Pheidole*	sp. 14			W
Myrmicinae	*Pheidole*	sp. 15	W		W
Myrmicinae	*Pheidole*	sp. 16			W
Myrmicinae	*Pheidole*	sp. 17		W	
Myrmicinae	*Pheidole*	sp. 18	W		
Myrmicinae	*Pheidole*	sp. 19		W	
Myrmicinae	*Pheidole*	sp. 20	W		
Myrmicinae	*Pheidole*	sp. 21	W		W
Myrmicinae	*Pheidole*	sp. 22	H		
Myrmicinae	*Pheidole*	sp. 23	W		
Myrmicinae	*Pheidole*	sp. 24			W
Myrmicinae	*Pheidole*	sp. 25	W		

table continued on next page

Appendix 7.1. continued

Subfamily	Genus	Species	Upper Palumeu (270 m) 5 days sampling, 2 winkler transects	Grensgebergte (800 m) 1 day sampling, 2 winkler transects*	Kasikasima (200 m) 6 days sampling, 3 winkler transects
Myrmicinae	*Pheidole*	sp. 26	W		
Myrmicinae	*Pheidole*	sp. 27			W
Myrmicinae	*Pheidole*	sp. 28		H	
Myrmicinae	*Pheidole*	sp. 29	H		
Myrmicinae	*Pheidole*	sp. 30	W		
Myrmicinae	*Pheidole*	sp. 31	W		W
Myrmicinae	*Pheidole*	sp. 32 (cf. *minutula*)	H		H
Myrmicinae	*Pheidole*	sp. 33	W		
Myrmicinae	*Pheidole*	sp. 34	W		
Myrmicinae	*Pheidole*	sp. 35	W		
Myrmicinae	*Pheidole*	sp. 36	W		
Myrmicinae	*Pheidole*	sp. 37	W		
Myrmicinae	*Rogeria*	sp. 1			W
Myrmicinae	*Rogeria*	sp. 2	W		
Myrmicinae	*Solenopsis*	sp. 1	W		W
Myrmicinae	*Solenopsis*	sp. 2			W
Myrmicinae	*Solenopsis*	sp. 3	W		W
Myrmicinae	*Solenopsis*	sp. 4	W	W	W
Myrmicinae	*Solenopsis*	sp. 5			W
Myrmicinae	*Solenopsis*	sp. 6	W		W
Myrmicinae	*Solenopsis*	sp. 7			W
Myrmicinae	*Solenopsis*	sp. 8	W		W
Myrmicinae	*Solenopsis*	sp. 9		W	
Myrmicinae	*Solenopsis*	sp. 10	W	W	
Myrmicinae	*Solenopsis*	sp. 11 (fire ant)		H	
Myrmicinae	*Solenopsis*	sp. 12 (fire ant)	H		
Myrmicinae	*Strumigenys*	cf. *subdentata*	W		W
Myrmicinae	*Strumigenys*	*denticulata*			W
Myrmicinae	*Strumigenys*	*perparva*			W
Myrmicinae	*Strumigenys*	*beebei*	W		W
Myrmicinae	*Strumigenys*	sp. 1	W		
Myrmicinae	*Trachymyrmex*	*ruthae*			W
Myrmicinae	*Trachymyrmex*	sp. 1	W		W
Myrmicinae	*Wasmannia*	*auropunctata*	W	W	W
Paraponerinae	*Paraponera*	*clavata*	H		H
Ponerinae	*Anochetus*	*horridus*	W		
Ponerinae	*Anochetus*	sp. 1			W
Ponerinae	*Anochetus*	sp. 2			W
Ponerinae	*Anochetus*	sp. 3			W

table continued on next page

Appendix 7.1. continued

Subfamily	Genus	Species	Upper Palumeu (270 m) 5 days sampling, 2 winkler transects	Grensgebergte (800 m) 1 day sampling, 2 winkler transects*	Kasikasima (200 m) 6 days sampling, 3 winkler transects
Ponerinae	*Hypoponera*	sp. 1		W	W
Ponerinae	*Hypoponera*	sp. 2			W
Ponerinae	*Hypoponera*	sp. 3			W
Ponerinae	*Hypoponera*	sp. 4	W		W
Ponerinae	*Hypoponera*	sp. 5			W
Ponerinae	*Hypoponera*	sp. 6			W
Ponerinae	*Hypoponera*	sp. 7			W
Ponerinae	*Hypoponera*	sp. 8	W		W
Ponerinae	*Hypoponera*	sp. 9		W	
Ponerinae	*Hypoponera*	sp. 10	W		W
Ponerinae	*Hypoponera*	sp. 11	W		
Ponerinae	*Hypoponera*	sp. 12	W		
Ponerinae	*Odontomachus*	cf. *bauri*			W
Ponerinae	*Odontomachus*	sp. 1			W
Ponerinae	*Odontomachus*	sp. 2		W, H	
Ponerinae	*Odontomachus*	sp. 3	H		
Ponerinae	*Pachycondyla*	sp. 1			W
Ponerinae	*Pachycondyla*	sp. 2			W
Ponerinae	*Pachycondyla*	sp. 3			W
Ponerinae	*Pachycondyla*	sp. 4			W
Ponerinae	*Pachycondyla*	sp. 5		W	
Ponerinae	*Pachycondyla*	sp. 6	H		W
Ponerinae	*Pachycondyla*	sp. 7	W		
Ponerinae	*Pachycondyla*	sp. 8	W		
Ponerinae	*Pachycondyla*	sp. 9	W		
Ponerinae	*Pachycondyla*	sp. 10		H	
Ponerinae	*Pachycondyla*	sp. 11		H	
Proceratiinae	*Discothyrea*	cf. *denticulata*			W
Pseudomyrmecinae	*Pseudomyrmex*	sp. 1		H	
Pseudomyrmecinae	*Pseudomyrmex*	sp. 2			H
Pseudomyrmecinae	*Pseudomyrmex*	sp. 3	W		
Number of Species collected at each RAP site			72	25	92
Number of Species in Winkler Samples			48	16	82
Number of Species from Hand Collecting (limited)			24	10	10 (4 from ant plants)
Total Number of Species collected at all 3 RAP sites			149		
Total Number of Species in Winkler Samples			118		
Total Number of Species from Hand Collecting			34		

*rainy conditions likely affected results of winkler transects (see text)

Chapter 8

Fishes of the Palumeu River, Suriname

Jan H. Mol and Kenneth Wan Tong You

SUMMARY

Eighteen sites near three camps along the Upper and Middle Palumeu River, Suriname, were sampled between 9 and 25 March, 2012. We recorded 94 species of fishes, and, in combination with a collection of fishes from Lower Palumeu River by Covain et al. in 2008, 128 species are now known to occur in Palumeu River. This diversity is high compared to the rest of the world, but is typical for a medium-sized river of the Guiana Shield. Alpha diversity was high in the Upper Palumeu River (71 species), while two sites in the Middle Palumeu River had low α diversity probably because they could not be assessed adequately due to high water level and strong current that reduced sampling effectiveness, especially in the rapids. We collected eleven species of fishes potentially new to science, including a *Bryconops* species with red fins, a small *Parotocinclus* catfish (*Parotocinclus* aff. *collinsae*), and a head-and-tail-light tetra (*Hemigrammus* aff. *ocellifer*). Of these eleven species, gen.nov. sp.n. aff. *Parotocinclus* was collected before in the Upper Marowijne River, but both this species and its genus are still undescribed. Two species are new records for Suriname: *Hyphessobrycon heterorhabdus* and *Laimosemion geayi*; a third and fourth species, *Ituglanis nebulosus* and *Pimelodella megalops*, may also represent new species for Suriname if their identity is confirmed. One species, *Aequidens paloemeuensis*, is endemic to Palumeu River, while an additional 6 species are endemic to the Marowijne River System, which includes Palumeu River: *Cyphocharax biocellatus*, *Semaprochilodus varii*, *Jupiaba maroniensis*, *Moenkhausia moisae*, gen.nov. sp.n. *Parotocinclus* and *Pimelodella procera*. An additional five species of Palumeu River collected by Covain et al. in 2008 are also endemic to the Marowijne River System: *Hemiodus huraulti*, *Corydoras* aff. *breei*, *Hemiancistrus medians*, *Pimelabditus moli* and *Platydoras* sp. We collected 71 species in Upper Palumeu River and tributaries (Site 1), 16 species in the Makrutu and Tapaje creeks (Site 3), and 49 species at the Kasikasima site (Site 4). The differences in the number of species among the three sites probably reflect differences in opportunity to sample effectively with seine net in shallow water at the three sites, diversity of habitat types that could be sampled at each site and total sampling time at each site. Species composition varied strongly among sites: sites 3 and 4 included large-sized fishes from the main channel of the Middle Palumeu River, while site 1 had many small-sized species of creek habitat. Overall, large top-level predators were still common in Palumeu River, indicating intact ecosystems. The primary threat to the fishes of Palumeu River is the so-called Tapajai Project, which proposes to build one or more dams in the Tapanahony River in order to divert its water via Jai Creek to Brokopondo Reservoir and thus increase power generation by the hydroelectric station at Afobaka. The dam(s) would directly affect migratory fishes, fishes of running water and creek habitats and fishes downstream of the dam(s). Several migratory fish species, which people throughout Suriname depend upon for food, may require the pristine headwaters of Southeastern Suriname for spawning, although very little is known. Furthermore, the dam(s) would effectively mix the fish faunas of the Marowijne River System and the Suriname River System, which each support distinct communities with endemic species, possibly leading to species declines or extinctions.

INTRODUCTION

Fishes are a critical source of protein to the Trio and Wayana people in villages along the Tapanahony and Palumeu rivers. They are a common and highly-valued component of many meals. Large and medium-sized fish species that are routinely eaten have local names. However, most fish species are small and often ignored by local people. However, many of these small species are highly valued in the aquarium hobby and could play an important role in the development of the area if fisheries for these species are strictly regulated. Fish species are an important component of the aquatic ecosystem. Large top-predators like anyumara (*Hoplias aimara*) (see page 19) and detritivores like kwimata (*Prochilodus* and *Semaprochilodus*) are often keystone species in aquatic ecosystems (Schindler 2007); as popular food fishes they are also most vulnerable to overfishing. Smaller fishes forage on aquatic invertebrates and serve as prey for larger fishes, caiman, and birds. Fish excretion can be an important source of recycled nutrients readily available to nutrient-starved primary producers in tropical streams (Vanni 2002, McIntyre et al. 2007). Recycling of limiting nutrients from prey or detritus pools represents one of the dominant sources of nutrients to aquatic primary producers in many ecosystems. Thus changes in species composition and fish diversity have the potential to alter the availability of limiting nutrients to primary producers, and hence ecosystem production, in systems where fish excretion is an important nutrient source (Taylor et al. 2006, McIntyre et al. 2007, Schindler 2007). Fish diversity reflects the health of the river systems.

The Palumeu River is a large tributary of the Tapanahony River, which itself is the largest tributary of the Marowijne (Maroni) River. There is a modest amount of published information concerning the fishes of the Marowijne River System (which includes Palumeu River). Most fish surveys in the Marowijne watershed are in the Marowijne / Lawa River proper or in the Litani, Tapanahony and Oelemari tributaries (e.g., Planquette et al. 1996, Ouboter et al. 1999). Other fish collections in the Upper Marowijne River have not been published but were included in reviews (e.g. Le Bail et al. 2012, Mol et al. 2012, Mol 2012), while the important 1966-collections of King Leopold III of Belgium from the Palumeu River (housed in Brussels) have yet to be studied. Le Bail et al. (2012) mention 242 strictly freshwater fish species and 37 marine cycle species (total 279 species) for the Marowijne (Maroni) River System, while Mol et al. (2012) have 314 fish species listed for the Marowijne River System. To our knowledge, there have been two scientific fish collections in the Palumeu River: the collection by Covain et al. in 2008 in the Lower Palumeu River (unpublished; Table 8.1) and the collection of King Leopold III of Belgium in 1966 in the Lower and Middle Palumeu River (collection yet to be studied; Table 8.2). This expedition was the first to collect fishes in the Upper Palumeu River, and will serve as a baseline for subsequent aquatic biodiversity studies.

STUDY SITES AND METHODS

Eighteen sites near three camps were sampled between 9 and 25 March 2012. Fishes were collected with a 3×2-meter fine-mesh seine, dip nets, 30-meter trammel nets, and hook and line. We also talked extensively with our Trio and Wayana guides about their knowledge of local fishes, and to discern what they were catching during the expedition while they were fishing for food. Every habitat was sampled with as many methods as practical in order to rapidly assess the diversity of the region and maximize the number of species observations. Rocks and woody debris were scraped. Submerged logs were cut open. Leaf litter was searched. Seines were pulled through patches of vegetation, leaf litter or submerged roots, as well as over sandy substrates. Canoes were used to travel extensively upstream and downstream from the camps. We also walked through the forest to survey

Table 8.1. Fish species collected in the period 27–31 October 2008 at Weyu Rapid and a small tributary forest creek upstream of Weyu Rapid, Lower Palumeu River, Suriname, by R. Covain, J. Montoya-Burgos, J.H. Mol and K. Wan Tong You. The forest creek was sampled with rotenone. Species marked with an asterisk were not collected during the present study.

Taxon	Weyu Rapid	Unnamed tributary of Lower Palumeu River upstream of Weyu Rapid
Characiformes		
Parodontidae		
Parodon guyanensis	x	
Curimatidae		
Cyphocharax spilurus		x
Cyphocharax sp.	x	
Prochilodontidae		
Prochilodus rubrotaeniatus*	x	
Semaprochilodus varii	x	
Anostomidae		
Anostomus brevior		x
Hypomasticus despaxi		x
Leporinus fasciatus	x	
Leporinus maculatus*	x	
Leporinus nijsseni*		x
Chilodontidae		
Caenotropus maculosus	x	
Crenuchidae		
Melanocharacidium sp.	x	
Hemiodontidae		
Hemiodus huraulti*	x	
Alestidae		
Chalceus macrolepidotus	x	
Characidae		
Bryconops affinis		x
Bryconops caudomaculatus		x
Bryconops sp.	x	
Jupiaba abramoides		x
Moenkhausia aff. *intermedia**	x	
Moenkhausia oligolepis		x
Moenkhausia sp.	x	x
Brycon pesu*	x	
Triportheus brachipomus	x	
Acnodon oligacanthus*	x	
Myloplus rhomboidalis	x	
Myloplus rubripinnis	x	
Myloplus sp.	x	
Pristobrycon eigenmanni	x	
Pristobrycon striolatus	x	

table continued on next page

Table 8.1. *continued*

Taxon	Weyu Rapid	Unnamed tributary of Lower Palumeu River upstream of Weyu Rapid
Serrasalmus rhombeus	x	
Tometes lebaili	x	
Cynopotamus essequibensis	x	
Roeboexodon guyanensis	x	
Bryconamericus guyanensis	x	
Tetragonopterus chalceus	x	
Acestrorhynchidae		
Acestrorhynchus falcatus		x
Erythrinidae		
Erythrinus erythrinus		x
*Hoplerythrinus unitaeniatus**		x
Hoplias aimara	x	x
Hoplias malabaricus		x
Lebiasinidae		
Pyrrhulina filamentosa		x
Siluriformes		
Cetopsidae		
Helogenes marmoratus		x
Trichomycteridae		
Ituglanis amazonicus		x
Callichthyidae		
Callichthys callichthys		x
Corydoras aff. *breei**	x	x
*Corydoras guianensis**		x
*Corydoras nanus**		x
Loricariidae		
Cteniloricaria platystoma	x	
Harttia guianensis	x	
*Metaloricaria paucidens**	x	
Rineloricaria sp. nov.*		x
*Ancistrus temminckii**	x	x
Guyanancistrus brevispinis	x	
*Hemiancistrus medians**	x	
*Hypostomus tapanahoniensis**	x	
*Lithoxus planquettei**	x	
*Pseudancistrus barbatus**	x	

Taxon	Weyu Rapid	Unnamed tributary of Lower Palumeu River upstream of Weyu Rapid
Pseudopimelodidae		
*Batrochoglanis raninus**	x	
*Pseudopimelodus bufonius**	x	
Heptapteridae		
*Imparfinis pijpersi**	x	
*Pimelodella cristata**		x
*Rhamdia quelen**		x
Pimelodidae		
*Pimelabditus moli**	x	
*Pimelodus ornatus**	x	
*Pseudoplatystoma fasciatum**	x	
Doradidae		
*Doras micropoeus**	x	
Platydoras sp.*	x	
Auchenipteridae		
*Auchenipterus dentatus**	x	
Ageneiosus inermis	x	
Gymnotiformes		
Gymnotidae		
*Electrophorus electricus**		x
Gymnotus carapo		x
Sternopygidae		
Sternopygus macrurus	x	
Hypopomidae		
Hypopomus artedi		x
Perciformes		
Sciaenidae		
*Pachypops fourcroi**	x	
Cichlidae		
*Cleithracara maronii**		x
*Crenicichla multispinosa**	x	
Crenicichla gr. *saxatilis*		x
Guianacara owroewefi	x	
*Krobia itanyi**	x	x
Total number of species = 79	**53**	**31**

Table 8.2. Fish collection localities and dates of King Leopold III of Belgium and J.P. Gosse in Palumeu River, Suriname, 25 October - 8 November 1966. These collections have yet to be studied.

Fieldnumber (Station)	Locality	Date
151	Tapanahoni River at Palumeu Village	25 October 1966
151a	Palumeu River at Kasikasima	27 October 1966
152	Palumeu River at Kasikasima	28 October 1966
153	Palumeu River at Papadron Rapid	1 November 1966
154	Waloemeroe Creek, right tributary of Palumeu River	2–3 November 1966
155	rapids of Palumeu River upstream of Waloemeroe Creek	3 November 1966
156	small right tributary of Palumeu River between Trombaka Noord Rapid and Trombaka Zuid Rapid	8 November 1966
157	Palumeu River at Trombaka Noord Rapid	8 November 1966

creeks, mountain brooks and a waterfall. Many individuals were released, but representative specimens were preserved in 4% formalin and later transferred to 70% ethanol for

Figure 8.1. A 50-m high waterfall in a right tributary of Upper Palumeu River was the collection site of a diverse fish fauna including many interesting characoids and loricariid and trichomycterid catfishes that are either new to science or collected for the first time in Suriname.

long-term storage at the National Museum of Natural History, Smithsonian Institution, in Washington, DC, USA.

For convenience sake we distinguish a lower section of Palumeu River (including Weyu Rapid), a middle section (including the large Tronbaka and Papadron rapid complexes and sites 3 and 4) and an upper headwater section (upstream of Tapaje Creek, site 1; strongly meandering and mostly without rapids). Fishes were collected in the Upper and Middle Palumeu River near three camp sites: first the headwaters of Palumeu River and tributary forest creeks at basecamp (site 1), secondly the large Makrutu and Tapaje tributaries, upstream of the Papadron Rapid (site 3), and finally the Middle Palumeu River in the reach between the Papadron and Tronbaka rapid complexes and some mountain brooks crossing the trail up the Kasikasima Mountain (site 4).

At site 1 the Upper Palumeu River is a narrow (10–15 m wide), shallow, and strongly meandering stream with characteristics of a medium-sized forest creek without rapids. The headwaters of Palumeu River have large amounts of woody debris (fallen trees) which make navigating the river difficult and time consuming. Downstream of basecamp, the inner bends of the river have remarkable herbaceous shore vegetation that is inundated during high water levels; upstream of basecamp, the river runs mostly under a closed forest canopy and is very shallow (mainly <2 m water depth). Considerable areas of riparian forest are flooded during high-water levels (e.g. our basecamp and trail to the helicopter pad). No people were seen, but apparently our camp site is also used by Trios from Brazil when they travel to Palumeu Village and Weyu Rapid. Fishes were collected in the main channel of the river with seine, trammel nets and hook-and-line, and in tributaries with seine nets. One right tributary of the Upper Palumeu River was characterized by a 50-m high waterfall (Figure 8.1) and with seine net we collected fishes both up- and downstream of this waterfall.

The Makrutu and Tapaje creeks (site 3) are rather large (70–100 m stream width) and deep (>2 m) tributaries of the Palumeu River which are better described as small rivers (at the confluence with Palumeu River they are both of approximately the same size as the Palumeu River itself); they were fished mainly with trammel nets and hook-and-line. The lower Makrutu Creek had extensive herbaceous shore vegetation of *Montrichardia* sp. (Sur. mokomoko) while Tapaje Creek had high dryland forest (terra firme) along its banks. A rapid complex and river island were situated immediately downstream of the confluence of Makrutu Creek and Palumeu River. No people were seen in the area. At this point along the river course the Palumeu River no longer has any true meanders and extensive flood plains, but instead the channel tends to follow troughs that coincide with joints and faults in the rock (thus irregular and angular bends).

Site 4 (Kasikasima) was situated in the middle reach of the Palumeu River, between the large Papadron and Tronbaka rapid complexes. There is a small village on the right bank of the Palumeu River, i.e. opposite our camp site, and a transit tourist camp (for the ascent of Kasikasima Mountain) approximately 2 km downstream of our camp. We fished in the main channel of the river with trammel nets and hook-and-line, but heavy rains created high water-level and strong current which limited sampling effectiveness in the rapids (see below). With seine nets we fished in a small lowland tributary near our camp and in three mountain brooks crossing the trail to Kasikasima.

RESULTS AND DISCUSSION

We recorded 94 species of fishes in the Upper and Middle Palumeu River (Appendix 8.1). This is typical for the interior of Suriname and nearby parts of the Guiana Shield; for example, the Coppename RAP survey recorded 112 species (Mol et al. 2006), the Sipaliwini/Kutari RAP survey recorded 99 species (Willink et al. 2011), the Oelemari River yielded 102 species (Ouboter et al. 1999), and a similar rapid assessment of the upper Essequibo River yielded 110 species (P.W. Willink et al. *unpublished data*). Although fish collection in the Upper Palumeu River and tributaries (site 1) was relatively effective, the Middle Palumeu River (sites 3 and 4) was certainly not adequately sampled due to high water levels in the river. Rapids and large tributaries (Makrutu and Tapaje creeks) have not been adequately sampled during the present study. A fish collection from the Lower Palumeu River in and near Weyu Rapid by Covain et al. (October 2008) (Table 8.1) yielded 34 fish species that were not collected during the present RAP survey: *Prochilodus rubrotaeniatus, Leporinus maculatus, Leporinus nijsseni, Hemiodus huraulti, Moenkhausia* aff. *intermedia, Brycon pesu, Acnodon oligacanthus, Hoplerythrinus unitaeniatus, Corydoras* aff. *breei, Corydoras guianensis, Corydoras nanus, Metaloricaria paucidens, Rineloricaria* sp.n., *Ancistrus temminckii, Hemiancistrus medians, Hypostomus tapanahoniensis, Lithoxus planquettei, Pseudancistrus barbatus, Batrochoglanis raninus, Pseudopimelodus bufonius, Imparfinis pijpersi, Pimelodella cristata, Rhamdia quelen, Pimelabditus moli, Pimelodus ornatus,*

Pseudoplatystoma fasciatum, Doras micropoeus, Platydoras sp., *Auchenipterus dentatus, Electrophorus electricus, Pachypops fourcroi, Cleithracara maronii, Crenicichla multispinosa,* and *Krobia itanyi.* Many, but not all, of these fish species live in rapids and most are expected to occur in the Middle Palumeu River. Combining the collections of October 2008 and the present RAP collection we arrive at a species list for Palumeu River of 128 fish species. This diversity is high compared to the rest of the world, but is typical for the Guiana Shield. Although the number of species collected in Palumeu River is thus comparable to the number of species collected in other rivers of the Guiana Shield, the species that were actually collected are only partly the same as those collected in, for example, Coppename River or Sipaliwini River. It is probably safe to state that all large rivers of the Guiana Shield have some fish species that are endemic to that specific river (see Le Bail et al. [2012] for French Guiana and Mol et al. [2012] for Suriname). The Palumeu River is part of the Marowijne River System and seven species we collected during the present study are endemic for that river system, but only one of these species, *Aequidens paloemeuensis,* is endemic to the Palumeu River proper (see below).

We collected eleven species of fishes potentially new to science, including a *Bryconops* species with red fins (photograph in Mol 2012 and page 22), a small *Parotocinclus* catfish with a *Microglanis*-like pigmentation pattern (*P.* aff. *collinsae,* Figure 8.2a), a head-and-tail-light tetra (*Hemigrammus* aff. *ocellifer*) (see page 23) with a black midlateral line running from the caudal spot to a vertical line through the beginning of the dorsal fin (photograph in Mol 2012). The other potentially new species are *Characidium* sp. (see page 22), *Ituglanis* sp., *Eigenmannia* sp. 2, *Rivulus* sp., *Jupiaba* sp., *Pimelodella* sp. (see page 22), gen.nov. sp.n. aff. *Parotocinclus,* and *Hyphessobrycon* sp. (heterorhabdus group) (see page 23). The *Eigenmannia* species (photograph in Mol 2012) was initially recognized on a previous RAP survey in the Coppename River (Willink and Sidlauskas 2006), and appears to be more widely distributed throughout Suriname's interior than originally thought (see Willink et al. 2011). The species gen.nov. sp.n. aff. *Parotocinclus* was collected before in the Upper Marowijne River (e.g. Le Bail et al. 2000), but both the species and its genus are still not described formally. A small ancistrine catfish with a striking dark dorsolateral pigmentation pattern on its head (collected at site 1 in a pool below a waterfall; Figure 8.1) was identified as postlarval *Guyanancistrus brevispinis* by S. Fisch-Muller (Museum National d'Histoire Naturelle, Genève, Switzerland) and thus not a new species. The number of eleven new species is typical for Neotropical rivers that have not been well surveyed by fish biologists.

Two species, *Hyphessobrycon heterorhabdus* (see page 23) and *Laimosemion geayi,* were new records for the country of Suriname (Mol et al. 2012; photographs in Mol 2012). *Laimosemion geayi* was previously only known from French Guiana and Brazil to the east of Suriname, while *H. heterorhabdus* was previously only known to occur in the Lower and Middle Amazon River to the south of Suriname (Géry 1977); however, *H. heterorhabdus* is also known from southwestern French Guiana (P.Y. Le Bail, pers. communication). Two additional species, *Ituglanis* cf. *nebulosus* (Fig. 8.2b) and *Pimelodella* cf. *megalops*, may also represent new records for Suriname if their identity can be confirmed; *Ituglanis nebulosus* is only known from its type locality in the Approuague River and Sinnamary River in French Guiana (De Pinna & Keith 2003, Le Bail et al. 2012) and we were not able to examine specimens of *I. nebulosus*; *Pimelodella megalops* is known from both Guyana (Eigenmann 1912) and French Guiana (Le Bail et al. 2012). We encountered seven fish species known only from the (Upper) Marowijne River System (i.e. endemic to the Marowijne River System): *Aequidens paloemeuensis, Cyphocharax biocellatus, Jupiaba maroniensis, Moenkhausia moisae*, gen.nov. sp.n. aff. *Parotocinclus* (see Le Bail et al. 2000, p. 262), *Pimelodella procera* and *Semaprochilodus varii*. Five other endemics of the Marowijne River System were collected by Covain et al. (2008) in the Palumeu River (Table 8.1): *Hemiodus huraulti, Corydoras* aff. *breei, Hemiancistrus medians, Pimelabditus moli* and *Platydoras* sp. The cichlid *A. paloemeuensis* seems restricted to its type locality and is the only species of these that is endemic to the Palumeu River proper.

We collected 71 species in Upper Palumeu River and tributaries (site 1), 16 species in the Makrutu and Tapaje creeks (site 3), and 49 species at the Kasikasima site (site 4) (Appendix 8.1). The differences in the number of species collected among the three sites probably reflect differences in opportunity to sample effectively with seine net in shallow water at the three sites, diversity of habitat types that could be sampled at each site and total sampling time at each site. The shallow (creek-like) Upper Palumeu River and its tributaries at site 1 were sampled easily with seine net and trammel nets in the main channel. The deep (river-like) Makrutu and Tapaje creeks (site 3) could only be sampled with trammel nets and hook-and-line, while site 4 included a few small creeks that could be sampled by seine besides the main channel of Middle Palumeu River that was sampled with trammel nets. Rapids that were present at both sites 3 and 4 could not be sampled effectively because of high water levels and strong currents.

The three sites also yielded different fish species. For example, site 1 yielded many small-sized fishes in the families Crenuchidae, Characidae, Trichomycteridae and Loricariidae that were not present at the other two sites, but the large predator *Hoplias aimara* was also common in the main channel of the meandering Upper Palumeu River. Especially interesting was the diverse fish fauna up- and immediately downstream a 50-m high waterfall in a right tributary of Upper Palumeu River (Fig. 8.1) which included several potentially new species or species collected for the first time in Suriname, e.g. gen.nov. sp.n aff. *Parotocinclus, Parotocinclus* aff. *collinsae, Bryconops* sp 'red fins', *Characidium* sp., *Hyphessobrycon heterorhabdus* and *Ituglanis* cf. *nebulosus*. Other species collected at this site are rare or have a restricted distribution: *Tetragonopterus rarus, Creagrutus melanzonus, Hemibrycon surinamensis, Bunocephalus aloikae*, *Lithoxus surinamensis*, and *Pimelodella procera*).

The main channel of the Middle Palumeu River at sites 3 and 4 yielded some large-sized species such as *Pseudoplatystoma tigrinum, Tometes lebaili, Semaprochilodus varii, Serrasalmus rhombeus* and *Brycon falcatus* (Fig. 8.3) that were absent from the headwaters. The tiger catfish *P. tigrinum* and the kwimata *S. varii* may be migratory species, while the pacu *Tometes lebaili* is endemic to the Marowijne River System. The mountain brooks crossing the trail to Kasikasima (site 4) yielded some small or medium-sized species characteristic of ephemeral streams: *Microcharacidium eleotrioides, Astyanax bimaculatus, Erythrinus erythrinus, Copella arnoldi, Pyrrhulina filamentosa, Callichthys callichthys, Helogenes marmoratus*, and three rivulines *Laimosemion geayi, Rivulus* aff. *holmiae* and *Rivulus* sp. (the latter potentially new to science).

Overall, large top-level predators like anyumara (*H. aimara*) were common in the Upper Palumeu River (site 1), an area that is difficult to access. Large piscivores were also present at sites 3 and 4, but at these sites their abundance was difficult to assess due to high water levels. In pristine environments, these types of fishes are abundant, but they are the first to disappear when there is excessive fishing pressure (Mol et al. 2006).

The primary threat to the fishes of the Palumeu River is the so-called Tapajai Project, which proposes to build one or more dams in the Middle Tapanahony River in order to divert its water via Jai Creek to Brokopondo Reservoir and thus increase power generation by the hydroelectric station at Afobaka. A recent study of the long-term impacts of the hydroelectric dam at Afobaka on the fish fauna of the Middle Suriname River (Mol et al. 2007) showed that out of

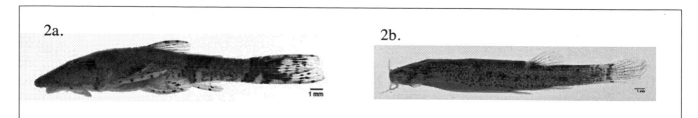

Figure 8.2. Two interesting fish species collected in a tributary of Upper Palumeu River. **A.** A potentially new *Parotocinclus* species (*P.* aff. *collinsae*) with a *Microglanis*-like pigmentation pattern. **B.** *Ituglanis* cf. *nebulosus* collected for the first time in Suriname. (photographs by Sandra Raredon, National Museum of Natural History, Smithsonian Institution)

172 species known from the Middle Suriname River before construction of the dam in 1964 only 41 fish species survived in Brokopondo Reservoir and that migratory species such as *Prochilodus rubrotaeniatus* no longer occurred downstream of the dam. Both *Prochilodus rubrotaeniatus* and the related *Semaprochilodus varii* occur in the Marowijne River System, which includes the Palumeu River. *Prochilodus* are keystone species in Neotropical rivers (Taylor et al. 2006), while *Semaprochilodus varii* is endemic to the Marowijne River system. Diversion of Tapanahony water to Brokopondo Reservoir would also (1) severely diminish the flow in the Marowijne River with consequent significant changes in aquatic habitats such as the area of flooded riparian forest and likely the fish fauna downstream of the reservoir and (2) effectively connect the Marowijne and Suriname river systems, each with their own endemic species (Mol et al. 2012). Mixing of these faunas may well lead to an ecological disaster, i.e. the introduction of Marowijne endemics into the Suriname River and *vice versa*, because changes in species interactions could lead to massive species declines or extinctions. We are unaware of any imminent plans for logging or (gold) mining in the area, but both would have negative impacts on the fish fauna by increasing erosion and sedimentation (which would affect bottom fishes and visually oriented fishes) and by decreasing the amounts of food that falls into the water, especially when the river floods. Gold mining in tributaries of the Upper and Middle Marowijne River and in the Marowijne River proper already has substantially increased turbidity from suspended sediments in the main channel: on 5 February 2012 we measured a Secchi disc transparency of only 5–10 cm in the Marowijne River near Nassau Mountains (km 105) (in 1990, Secchi transparency was >150 cm in the then still undisturbed Marowijne River). The increased turbidity (decreased transparency) in the main channel may affect long distance fish migrations in the Marowijne River and thus potentially also the migratory fishes of Palumeu River.

CONSERVATION RECOMMENDATIONS

Local communities along the Tapanahony River (including Palumeu Village) should be extensively informed about the potential impacts of the Tapajai Project on their immediate environment (including the fish fauna) so they can make rational, well-informed choices about their future with or without the Tapajai Project. The leaders of the villages should be given the opportunity to visit Brokopondo Reservoir and talk with elderly maroons who used to live in villages along the Middle Suriname River before construction of the Afobaka Dam in 1964 (and who now often live in transmigration villages, e.g. Brownsweg). Conserving the headwaters and middle reaches of the Palumeu River will be important for maintaining food for people for many years to come. Protecting forests and rivers of the upper Palumeu and other nearby headwaters in Southeastern Suriname will not only maintain the high water quality that sustains fish for people to eat downstream, but may also sustain the spawning grounds for migratory species that people depend upon. However, more research is needed to understand migratory fish behavior in Suriname.

The fishes of Palumeu River can be of interest to the aquarium hobby (*Hyphessobrycon roseus*, *Hyphessobrycon heterorhabdus*, *Hypomasticus despaxi*, *Anostomus brevior*, *Corydoras* spp., *Farlowella reticulata*, *Bunocephalus aloikae*, *Laimosemion geayi*, and many others) and sport fishers (*Hoplias aimara*, *Pseudoplatystoma tigrinum*, *Tometes lebaili*, *Brycon falcatus*) and thus generate income for local people if catches are regulated. Sustainable catch and export of aquarium fishes of the Palumeu River should be promoted.

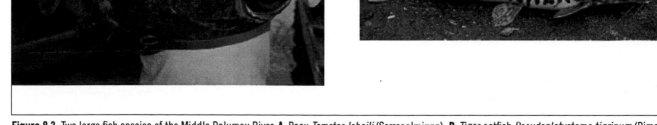

Figure 8.3. Two large fish species of the Middle Palumeu River. **A.** Pacu *Tometes lebaili* (Serrasalminae). **B.** Tiger catfish *Pseudoplatystoma tigrinum* (Pimelodidae).

Additional actions that should be taken:

1. Assess which fish species from the Palumeu River are used for food.
2. Determine amount caught and eaten.
3. Study life-history of these species to determine how fast they reproduce and grow.
4. Determine the amount of fish that can be sustainably harvested, both for food fishes and aquarium fishes.
5. Set catch limits and/or seasons if necessary to avoid overfishing.

In addition, it is recommended to create picture guides of fishes, especially colorful species and fun-to-catch species, in order to increase appreciation and knowledge of fishes. These guides can be used to promote ecotourism. In collaboration with METS ecotourism organization, we suggest setting up and maintaining a few aquariums in Palumeu Village with fishes of the river for the benefit of both local school children and ecotourists. Both guides and aquariums will increase understanding of the fish fauna of Palumeu River and thus promote conservation of the aquatic habitat and the fish fauna.

The collection of eleven potentially new fish species in the Upper and Middle Palumeu River under unfavorable (high-water) conditions during the present study and the absence of many fish species from the rapids in the present collection (as is clear from a comparison of the present collection with that of Covain et al. in October 2008) both indicate a richer fish fauna in Palumeu River than the fish fauna that is currently known (i.e. 128 species from the RAP collection and the collection of Covain et al 2008). In order to arrive at a more complete list of the fish fauna of Palumeu River we recommend:

Additional scientific surveys are necessary to document the fish biodiversity. There are species present that we did not collect (see for example Table 8.1), and there could be new species to science yet to be discovered. Additional surveys should be conducted at different times of the year, but especially when river levels are lower; and collection efforts should be aimed mainly at the major rapid complexes and the main river channel and tributaries in middle reaches of the Palumeu River.

The 1966-collections of King Leopold III of Belgium and J.P Gosse in Palumeu River (Table 8.2) should be studied. These collections, which include collections in the large Trombaka and Papadron rapid complexes during the dry (low-water) season (27 October – 8 November 1966), are housed in Brussels and have not been studied as far as we know (Richard Vari, Smithsonian Institution, American Museum of Natural History, *personal communication*).

REFERENCES

Eigenmann, C.H. 1912. The freshwater fishes of British Guiana, including a study of the ecological grouping of species and the relation of the fauna of the plateau to that of the lowlands. Memoirs Carnegie Museum. 5: 1–578.

Géry, J. 1977. Characoids of the world. TFH Publications, Neptune City, USA.

Le Bail, P.Y., P. Keith, and P. Planquette. 2000. Atlas des poissons d'eau douce de Guyane. Tome 2 – fascicule II. Siluriformes. Museum National d'Histoire Naturelle, Paris.

Le Bail, P.Y., R. Covain, M. Jégu, S. Fisch-Muller, R. Vigouroux, and P. Keith. 2012. Updated checklist of the freshwater and estuarine fishes of French Guiana. Cybium. 36: 293–319.

McIntyre, P.B., L.E. Jones, A.S. Flecker, and M.J. Vanni. 2007. Fish extinctions alter nutrient recycling in tropical freshwaters. Proceedings of the National Academy of Sciences USA. 104: 4461–4466.

Mol, J.H. 2012. The freshwater fishes of Suriname. Brill, Leiden.

Mol, J.H., P. Willink, B. Chernoff, and M. Cooperman. 2006. Fishes of the Coppename River, Central Suriname Nature Reserve, Suriname. *In*: Alonso, L.E. and H.J. Berrenstein (eds). A Rapid Biological Assessment of the Aquatic Ecosystems of the Coppename River Basin, Suriname. RAP Bulletin of Biological Assessment 39. Washington, DC: Conservation International. pp. 67–79.

Mol, J.H., B. de Mérona, P.E. Ouboter, and S. Sahdew. 2007. The fish fauna of Brokopondo Reservoir, Suriname, during 40 years of impoundment. Neotropical Ichthyology. 5: 351–368.

Mol, J.H., R.P. Vari, R. Covain, P.W. Willink, and S. Fisch-Muller. 2012. Annotated checklist of the freshwater fishes of Suriname. Cybium. 36: 263–292.

Ouboter, P.E., B.P.E. De Dijn, and U.P.D. Raghoenandan. 1999. Biodiversity inventory Ulemari Area: preliminary technical report. National Zoological Collection of Suriname & National Herbarium of Suriname, Paramaribo.

de Pinna, M. C. C., and P. Keith. 2003. A new species of the catfish genus *Ituglanis* from French Guyana (Osteichthyes: Siluriformes: Trichomycteridae). Proceedings of the Biological Society of Washington. 116: 873–882.

Planquette, P., P. Keith, and P.Y. Le Bail. 1996. Atlas des poissons d'eau douce de Guyane. Tome 1. Museum National d'Histoire Naturelle, Paris.

Taylor, B.W., A.S. Flecker, and R.O. Hall. 2006. Loss of a harvested fish species disrupts carbon flow in a diverse tropical river. Science. 313: 833–836.

Schindler, D.E. 2007. Fish extinctions and ecosystem functioning in tropical ecosystems. Proceedings of the National Academy of Sciences USA. 104: 5707–5708.

Vanni, M.J. 2002. Nutrient cycling by animals in freshwater ecosystems. Annual Review Ecology Systematics. 33: 341–370.

Willink, P.W. and B.L. Sidlauskas. 2006. Taxonomic notes on select fishes collected during the 2004 AquaRAP expedition to the Coppename River, Central Suriname Nature Reserve, Suriname. *In*: Alonso, L.E. and H.J. Berrenstein (eds). A Rapid Biological Assessment of the Aquatic Ecosystems of the Coppename River Basin, Suriname. RAP Bulletin of Biological Assessment 39. Washington, DC: Conservation International. pp. 101–111.

Willink, P.W., K. Wan Tong You, and M. Piqué. 2011. Fishes of the Sipaliwini and Kutari Rivers, Suriname. *In*: O'Shea, B.J., L.E. Alonso, and T.H. Larsen (eds). A Rapid Biological Assessment of the Kwamalasamutu region, Southwestern Suriname. RAP Bulletin of Biological Assessment 63. Washington, DC: Conservation International. pp. 118–123.

Appendix 8.1. List of fish species recorded in the Upper and Middle Palumeu River, Suriname.

Taxon	Upper Palumeu River (basecamp, site 1) n = 9	Makrututu and Tapaje Creeks (Site 3) n = 3	Middle Palumeu River and mountain streams (Site 4) n = 6
Characiformes			
Parodontidae			
Parodon guyanensis	3	2	0
Curimatidae			
Cyphocharax biocellatus	8	0	5
Cyphocharax helleri	3	0	10
Cyphocharax spilurus	0	0	1
Prochilodontidae			
Semaprochilodus varii	0	0	1
Anostomidae			
Anostomus brevior	16	0	2
Hypomasticus despaxi	1	0	14
Leporinus fasciatus	0	0	1
Leporinus friderici	4	4	0
Leporinus gossei	0	0	1
Chilodontidae			
Caenotropus maculosus	1	0	0
Crenuchidae			
Characidium zebra	24	0	0
Characidium sp. A	13	0	0
Melanocharacidium blennioides	10	0	0
Melanocharacidium dispilomma	13	0	0
Microcharacidium eletrioides	0	0	6
Hemiodontidae			
Bivibranchia bimaculata	0	0	3
Alestidae			
Chalceus macrolepidotus	7	0	0
Characidae			
Astyanax bimaculatus	0	0	1
Bryconops affinis	4	0	0
Bryconops caudomaculatus	37	4	0
Bryconops melanurus	2	0	28
Bryconops sp. (red fins)	35	0	0
Hemigrammus aff. *ocellifer*	18	0	3
Hyphessobrycon heterorhabdus	15	0	0
Hyphessobrycon sp. (heterorhabdus grp.)	0	0	2
Hyphessobrycon roseus	82	0	12
Jupiaba abramoides	7	0	0
Jupiaba keithi	6	0	2
Jupiaba maroniensis	1	3	1
Jupiaba sp. A	3	0	0
Moenkhausia georgiae	1	0	0
Moenkhausia grandisquamis	0	0	1
Moenkhausia inrai	0	0	7

table continued on next page

Appendix 8.1. *continued*

Taxon	Upper Palumeu River (basecamp, site 1) n = 9	Makrututu and Tapaje Creeks (Site 3) n = 3	Middle Palumeu River and mountain streams (Site 4) n = 6
Moenkhausia moisae	9	0	4
Moenkhausia oligolepis	9	0	3
Brycon falcatus	0	1	0
Triportheus brachypomus	3	0	1
Myloplus rhomboidalis	0	1	2
Myloplus rubripinnis	2	4	2
Myloplus ternetzi	14	0	2
Mylopus sp (juvenile)	8	1	0
Pristobrycon eigemanni	62	8	2
Pristobrycon striolatus	0	4	3
Serrasalmus rhombeus	0	24	1
Tometus lebaili	0	0	1
Cynapotamus essequibensis	18	0	0
Phenacogaster wayana	79	0	3
Roeboexodon guyanensis	1	0	2
Poptella brevispina	12	0	0
Tetragonopterus chalceus	8	3	4
Tetragonopterus rarus	3	0	0
Bryconamericus guyanensis	0	1	0
Creagrutus melanzonus	52	0	0
Hemibrycon surinamensis	10	0	0
Acestrorhynchidae			
Acestrorhynchus falcatus	2	0	0
Acestrorhynchus microlepis	4	0	1
Erythrinidae			
Erythrinus erythrinus	0	0	1
Hoplias aimara	8	1	3
Hoplias malabaricus	1	0	0
Hoplias sp (juvenile)	3	0	0
Lebiasinidae			
Copella arnoldi	4	0	14
Nannostomus bifasciatus	23	0	4
Pyrrhulina filamentosa	0	0	2
Siluriformes			
Cetopsidae			
Helogenes marmoratus	0	0	1
Aspredinidae			
Bunocephalus aloikae	4	0	0
Trichomycteridae			
Ituglanis amazonicus	4	0	0
Ituglanus cf. *nebulosus*	15	0	0
Ituglanis sp. A	1	0	0
Ochmacanthus sp.	1	0	0

table continued on next page

Appendix 8.1. *continued*

Taxon	Upper Palumeu River (basecamp, site 1) n = 9	Makrututu and Tapaje Creeks (Site 3) n = 3	Middle Palumeu River and mountain streams (Site 4) n = 6
Callichthyidae			
Callichthys callichthys	1	0	1
Loricariidae			
Otocinclus mariae	3	0	0
Gen.nov. sp.n aff. *Parotocinclus*	34	1	0
Parotocinclus aff. *collinsae*	3	0	0
Cteniloricaria platystoma	1	0	0
Farlowella reticulata	3	0	0
Harttia guianensis	9	0	0
Ancistrus aff. *hoplogenys*	13	0	0
Guyanancistrus brevispinis	8	0	0
Lithoxus surinamensis	10	0	0
Pseudopimelodidae			
Microglanis poecilus	32	0	0
Heptapteridae			
Pimelodella cf. *megalops*	1	0	0
Pimelodella procera	12	0	0
Pimelodella sp. A	4	0	0
Pimelodidae			
Pseudoplatystoma tigrinum	0	2	0
Auchenipteridae			
Ageneiosus inermis	52	5	2
Gymnotidae			
Gymnotus carapo	1	0	1
Sternopygidae			
Sternopygus macrurus	1	0	0
Eigenmannia sp. 2	1	0	0
Hypopomidae			
Hypopomus artedi	2	0	1
Perciformes			
Cichlidae			
Aequidens paloemeuensis	4	0	1
Crenicichla albopunctata	14	0	0
Crenicichla sp (juvenile)	2	0	1
Guianacara owroewefi	60	0	27
Cyprinodontiformes			
Rivulidae			
Laimosemion geayi	0	0	7
Rivulus aff. *holmiae*	0	0	4
Rivulus sp. A	0	0	4
Total 94 species	71 species	16 species	49 species

Chapter 9

A herpetofaunal survey of the Grensgebergte and Kasikasima regions, Suriname

Stuart Nielsen, Rawien Jairam, Paul Ouboter and Brice Noonan

SUMMARY

We conducted a herpetofaunal inventory at four sites in Southeastern Suriname from March 8–28th 2012, and recorded 47 species of amphibians and 42 species of reptiles. These numbers are lower than other areas within the Guiana Shield that are better sampled (e.g. Iwokrama, Guyana; Nouragues, French Guiana), but are relatively high when compared with other sites sampled over the same time period (e.g., recent RAP surveys in Suriname). Seven (six frogs and one snake) of the total 89 species encountered could not be assigned to any nominal species. These unidentified taxa may represent novel species, yet require validating genetic and morphological data before formal diagnoses can be made. A number of records represent range expansions for taxa within the Guiana Shield (e.g. *Rhinatrema bivitattum, Alopoglossus buckleyi*). Additionally, a teiid lizard (*Cercosaura argulus*) is recorded for just the second time in Suriname. Encountering >80 total species (including 19 snake species) is evidence of a healthy, diverse and seemingly pristine forest ecosystem.

INTRODUCTION

Reptiles and amphibians form a prominent, speciose component of tropical forests and many aspects of their biology (e.g. small body size in concert with large population sizes, intermediate roles in food webs, strict micro-habitat requirements, etc.) contribute to their value as a focal group for biotic surveys. Amphibians are very good indicators of disturbance (Stuart et al. 2004) because they are sensitive to changes in microclimate, particularly as most possess a biphasic lifestyle (i.e. two distinct life stages, larval and adult) heavily dependent on high quality water resources. Amphibians are well suited for rapid assessments as they are often easy to sample; but when that is not the case, their species-specific diagnostic calls aid passive identification, particularly for hard to collect species (e.g. canopy dwellers; Marty and Gaucher 2000). Biotic surveys of amphibians in particular are imperative as widespread and poorly understood disease vectors (e.g. chytrid fungus and ranavirus) are causing worldwide declines, even in seemingly pristine areas (Lips 1998). Lizards are more diverse in primary forest, compared to secondary or modified forest (i.e. plantation; Gardner et al. 2007), suggesting they are also sensitive to changes in microhabitat. Presence of turtles and tortoises can also be a good indicator of hunting pressure as they are often targeted for subsistence hunting by local Amerindians (Peres 2001). Although one of the smallest South American countries, Suriname possesses a wide variety of amphibians (>100 species according to Señaris and MacCullough 2005; 107 species according to Ouboter and Jairam 2012) and reptiles (>170 species; Ávila Pires 2005). While very few of these species are endemic to Suriname itself, most are endemic to the larger Guiana Shield or the more inclusive Amazo-Guianan Subregion. The goal of this RAP survey in southern Suriname was to provide baseline information on the diversity and abundance of amphibians and reptiles for the areas in and around the Grensgebergte and Kasikasima Mountains. We sampled four sites incorporating both upland and lowland habitat, from seasonally flooded forest to human modified secondary forest to exposed granite outcrops. We also provide basic statistics comparing our findings with other RAP surveys in Suriname, as well as other well-studied regions in the Guianas (e.g. Iwokrama, Guyana; Nouragues, French Guiana). Finally, we discuss conservation recommendations for the region.

METHODS

Of the four main RAP study sites, herpetological collections were made in only three (Upper Palumeu River—Site 1 [9 days], Grensgebergte Mountains—Site 2 [2 days], and Kasikasima—Site 4 [6 days]; the unsampled site (Site 3) was visited only by the aquatic team while they were heading downriver between Sites 1 to 4). In addition, some species were encountered at the METS resort in Palumeu (Site 5 [1 day]), a subset of which was encountered at other sites.

In order to encounter as many species as possible, opportunistic encounters and captures were made primarily via active searching. We walked pre-established trails through forest, in forest clearings, and on stream banks. Surveys were conducted at various times of the day (although focusing on main activity periods at dusk and dawn) and early evening to mid-night. Passive frog call surveys were performed at night and if the calling male was suspected of being near the ground, efforts were made to locate it. Special attention was taken to search in a variety of habitats, particularly in streams/creeks and under logs/fallen bark—areas known to harbor rarely seen species or those with strict habitat requirements, as well as after inclement weather (i.e. heavy rains). We turned stones and logs, opened rotting logs, and raked litter to reveal hidden animals. At times, we deviated off of the main trail/transect to search adjacent habitat or features (e.g. fallen log, swampy pool) we deemed of interest or where animal activity was observed. Opportunistic surveys are effective for collecting as many species as possible in a short time period and sampling total species richness (Donnelly et al. 2004). Unfortunately, care was not taken to painstakingly record each individual of each species encountered, so only qualitative assessments of species relative abundance are provided (Appendix 9.1).

When possible and/or necessary, animals were caught either by hand, noose, net or rubber band, and preliminary identifications were made. Additionally, a portion of the total catch was euthanized (via subcutaneous injection of dilute Euthanaze®), fixed in 10% formalin solution, and then stored in 70% ethanol as museum voucher specimens (stored at the National Zoological Collection of Suriname, Anton de Kom, Universiteit van Suriname). These specimens were given unique field identification tags and many have representative life photographs (taken by S. Nielsen, P. Naskrecki and/or T. Larsen). R. Jairam later performed more rigorous, museum-based, morphological identification to verify species IDs. Samples of liver/muscle tissue for DNA analyses were extracted from voucher specimens (before formalin fixation) and stored in 95% ethanol (stored in the University of Mississippi frozen tissue collection).

We compared data on amphibian and reptile surveys from five sites in the Guiana Shield (Nouragues and Arataye, French Guiana—Born and Gaucher 2001; Petit Saut, French Guiana—Duellman 1997; Piste Ste. Elie, French Guiana—Born and Gaucher 2001; and Iwokrama, Guyana—Donnelly et al. 2005) originally assembled by Watling and Ngadino (2007), as well as three additional sites from two previous RAP surveys in Suriname (RAP 43, Watling and Ngadino 2007; RAP 63, Ouboter et al. 2011). Although undetected and/or undescribed species certainly exist throughout the Guiana Shield, the non-RAP surveys—which occurred over multiple seasons—are better sampled compared to RAP surveys (which generally span just a number of days), therefore, direct comparisons must be viewed with some uncertainty. Additionally, the geological complexity of the Guiana Shield makes comparisons of species communities/composition between high elevation inselbergs and low elevation seasonally flooded forests

difficult, as different habitats support different species and different overall levels of species richness may be expected may provide counterintuitive arguments. Instead, we provide these data as a benchmark with which to compare this current study.

RESULTS AND DISCUSSION

We observed a total of 89 species of reptiles and amphibians during this survey (Table 9.1). While most of the species are confidently sorted to known species, seven (six frogs and one snake) of the total species encountered could not be assigned to any nominal species and 23 are listed as *cf.*, meaning only informal identifications are possible without more comparative material and rigorous morphological and/or molecular examination. Appendix 9.1 lists the species we recorded and a qualitative measure of abundance for each species for the four collection localities. Upper Palumeu River (Site 1) provided the most diverse list of species (56 species, 63% of total—30 amphibians and 26 reptiles; of which, 12 amphibians and 16 reptiles were only encountered at Site 1; see Table 9.2). This result may partially be explained by the amount of time spent collecting in that locality versus the other sites (8 days total versus the 2nd longest stay—6 days [Site 4]), thus affording us greater opportunity to encounter a wider variety of species. However, some other subtle

Table 9.1. Herpetofaunal richness at 12 sites in the Guiana Shield, including data from the two previous RAP surveys to Nassau/Lely Mountains and Kwamalasamutu. In each column, data are presented as raw species number/percentage of total herpetofauna.

Site	Amphibians	Reptiles	Total
Iwokrama	37/0.34	71/0.66	108
Nourague	51/0.47	58/0.53	109
Arataya	62/0.49	65/0.51	127
Piste Ste. Elie	33/0.38	53/0.62	86
Brownsberg	64/0.44	80/0.56	144
Mean=	*49*	*65*	*115*
Nassau	16/0.55	13/0.45	29
Lely	20/0.55	16/0.45	36
Kwamalasamutu	42/0.54	36/0.46	78
Mean=	*26*	*22*	*48*
Upper Palumeu	30/0.54	26/0.46	56
Grensgebergte	6/0.46	7/0.54	13
Kasikasima	24/0.53	21/0.47	45
Palumeu	13/0.72	5/0.28	18
Mean=	*18*	*15*	*33*
Total species recorded on this RAP	47/0.53	42/0.47	89

Table 9.2. Breakdown of reptile and amphibian species encountered at each locality and the site-specific percentage of the total species recorded, as well as how unique each site was for both taxonomic groups.

Collection Site	1. Upper Pal.	2. Grensgebergte	4. Kasikasima	5. Palumeu
# of reptiles and amphibian species encountered (% of total sp. encountered [89 spp.])	56 (63%)	13 (15%)	45 (51%)	18 (20%)
# of amphibian species encountered (% of total amphibs [47 spp.])	30 (64%)	6 (13%)	24 (51%)	13 (28%)
# of amphib. species encountered that were unique to locality (% of spp. encountered for each site that were unique)	12 (40%)	1 (16%)	6 (25%)	7 (53%)
# of reptile species encountered (% of total reptile species encountered [42 spp.])	26 (62%)	7 (17%)	21 (50%)	5 (12%)
# of reptile species encountered that were unique (% of spp. encountered for each site that were unique)	16 (62%)	3 (43%)	10 (48%)	2 (40%)

vegetation or habitat differences might also have played a role (e.g. presence of seasonally flooded lowland forest in Site 1). Grensgebergte Mountain (Site 2: 13 species, 15% of total species encountered; 6 amphibians and 7 reptiles, of which 1 amphibian and 3 reptiles were only encountered at Site 2) was by no means speciose, but provided two unique snake records: *Dipsas copei* and the only sighting of *Bothrops atrox*. Kasikasima (Site 4) was the second most diverse site (45 species, 51% of total—24 amphibians and 21 reptiles; of which, 6 amphibians and 10 reptiles were encountered only at Site 4). Palumeu (Site 5) was not particularly diverse (18 species, 20% of total—13 amphibians and 5 reptiles; of which 7 amphibians and 2 reptiles were encountered only at Site 5), although we suspect that the surrounding area could potentially harbor other species. As the expedition was not focused on this area, the species encountered were all observed/collected within one 24-hr. period (including one rainy night) before continuing on to Site 1.

When the results for all localities are combined, our findings are comparable to those of RAP 63 of the Kwamalasamutu region of southwestern Suriname (Ouboter et al. 2011; see Table 9.1), which recorded 78 species of amphibians and reptiles (42 and 36, respectively), and is from a geographically close region. However, when each site is analyzed separately, our numbers are much more similar to those found by Watling and Ngadino (2007; RAP 43) in eastern Suriname, which also spent a similar amount of time sampling each of their two sites (~5 days). RAP 43 focused solely on areas of intermediate/high elevation, which harbored fewer (albeit more highland endemic) species. All recent RAP surveys, including the present study, generally recorded fewer species than other areas within the Guiana Shield (see Table 9.2), which can likely be explained by total days of search effort.

A number of species (24 of 89 total sp.) were recorded at the two sites that received the greater proportion of our search effort (Sites 1 & 4). Of the species that were recorded at just one site (45 spp.), most had congeneric relatives present at other sites, possibly/effectively filling similar ecological niche space (e.g. the snake *Atractus torquatus* was present only at Site 1 and *A. flammigerus* only at Site 4; however, one could argue that these two species occupy similar niches

in the ecosystem). Albeit an extraneous assumption, there remains the possibility that given more search effort, other congeneric (and potentially ecologically similar) species could also be found (or that competitive niche exclusion restricts them to microhabitats that we failed to survey adequately). This pattern is also seen when comparing our results to the two other recent RAP surveys (RAPs 43 & 63); of the 89 species recorded in the present study, RAP 43 (49 spp. total) recorded 23 (26%) and RAP 63 (78 spp. total) recorded 44 (49%) of the same species (see Appendix 9.1 and Table 9.1). However, in both of these two related surveys, congeneric species (which have the potential to be ecologically similar) were collected (see species marked as "N*" in Appendix 9.1). Thus, at a higher taxonomic scale, our results are similar (see further description below). A number of species that were recorded by these previous RAPs were noticeably absent from our collection efforts (e.g. frogs such as the microhylid, *Chiasmocleis shudikarensis,* the ceratophryid, *Ceratophrys cornuta*, and the pipid, *Pipa aspera*), although these species are generally less likely to be sampled due to their cryptic, semi-fossorial, and/or fully aquatic lifestyles.

The ~104 currently recognized amphibian species recognized in Suriname are representative of 38 genera in 13 families (Señaris and MacCullough 2005, Ouboter and Jairam 2012). Due to the "rapid," abbreviated nature of RAP surveys, we were unlikely to encounter all the biodiversity a geographic area harbors (especially as some of Suriname's ~104 amphibian species are restricted to coastal lowlands or other areas/habitat we were unlikely to survey in southern Suriname; see Ouboter and Jairam 2012). During the course of this RAP, we recorded ~47 species from 19 genera and 7 families (45%, 50%, and 54% of Suriname's currently recognized totals, respectively). Considering the aforementioned shortcomings of this type of survey, capturing roughly 50% of the known amphibian biodiversity of Suriname is a positive result. By comparison, RAP 43 (36 spp. from 13 genera and 5 families) and RAP 63 (42 spp. from 19 genera and 10 families) each recorded approximately 35% and ~50–70% of the diversity, respectively.

Although we observed and/or collected a smaller percentage of the total reptile diversity of Suriname (~170 spp., 92 gen., 23 fam.), our results are roughly similar in pattern to those for amphibians outlined above. We collected 42 species from 37 genera and 17 families (25%, 40% and 74% of the Surinamese total, respectively). Similar to the above results, RAP 43 collected a smaller percentage of the total (17% of Suriname's total spp., 24% of gen., 48% of fam.), whereas RAP 63 had results similar to (albeit less than) the present study (21% of Suriname's total spp., 35% of gen., 65% of fam.). Why these results represent a larger proportion of the higher-level diversity than we recovered in amphibians is unknown, but could be representative of an underestimate of amphibian familial diversity.

Without further work, it is difficult to say whether we adequately surveyed the region in our limited time during this survey, but we did achieve comparable results to recent, previous RAP surveys. Although the number of 'new' or rare species we collected was comparatively low (and really awaits thorough taxonomic revision of a number of groups) the real value of this region was the great diversity of herpetofauna seen and/or collected in a short time. We propose that these results suggest that southern Suriname is a local hotspot for herpetofaunal richness and if conserved in this pristine/semi-pristine state, this region will remain a true preserve of biodiversity.

Below are brief accounts of some of the species/taxonomic groups we encountered that are of interest due to their conservation status, distribution, natural history or potential as a new species, etc.

Class Amphibia, Order Anura

Allophrynidae: This family contains a single genus, *Allophryne*, and until this year was monotypic (Castroviejo-Fisher et al. 2012). The species we encountered, *A. ruthveni* is distributed throughout the Guiana Shield (GS; LaMarca et al. 2010). Numerous individuals were observed/captured at Site 1 that displayed a divergent color pattern when compared to "typical" *A. ruthveni*. As the recently described species from eastern Peru, *A. resplendens*, is diagnosed from *A. ruthveni* primarily on dorsal color pattern and mitochondrial DNA sequence divergence (Castroviejo-Fisher et al. 2012), we initially hoped we had found a third species for the genus. However, preliminary genetic data (unpub. 12S data) does not provide resounding evidence for elevating the Suriname population to a separate species. Further work is underway.

Dendrobatidae: We encountered two subfamilies within this super group of dart-poison frogs, Dendrobatinae and Aromobatinae. The latter contains >100 species in five genera, whereas Dendrobatinae contains 12 genera and >170 species. These frogs are terrestrial, largely diurnal, and display a unique reproductive mode exhibiting parental care.

Within Aromobatinae, we encountered four 'species' representative of two genera, *Allobates* and *Anomaloglossus*. *Allobates femoralis* is widely distributed across the Amazo-Guianan subregion (AGR; La Marca et al. 2010), whereas *Allobates granti* is distributed within a more limited area of the eastern GS (Kok 2008; Ouboter & Jairam 2012). Although neither species possessed particularly aberrant morphologies, Fouquet et al. (2012) has shown that there is considerable, geographically concordant genetic substructure across each species' respective range, potentially harboring complexes of cryptic species. As the taxonomic work is ongoing, it is difficult to accurately assess whether the populations sampled during this RAP survey are distinct from the larger Surinamese (or GS) groups delimited by Fouquet et al. (2012). *Anomaloglossus baeobatrachus* (IUCN Data Deficient) and *Anomaloglossus* sp. (see page 23) were abundant where they were found (Sites 1 and 4). Individuals of the latter found at the Kasikasima site possessed atypical (in comparison to *A. baeobatrachus*) dorsal color patterns (i.e. blotchy, not uniform). Similar to *Allobates*, Fouquet et al. (2012) has shown considerable, geographically concordant genetic divergence across these species' respective ranges, although it is currently unclear whether the *Anomaloglossus* "sp." referenced in that paper is the same taxon we encountered on this RAP. Unfortunately, at present we have been unsuccessful at obtaining informative DNA sequence information to compare to that of Fouquet et al. (2012), although work is ongoing.

Within Dendrobatinae, we encountered two species representative of two genera. Of the two species of *Amereega* that occur in Suriname, we found just *A. trivitatta* (see page 29), which is widespread across the AGR, although absent from the eastern GS. Such a wide distribution could harbor cryptic species, as has been found in closely related species within the genus (see Brown and Twomey 2009). However, of the individuals we collected, there was no significant morphological divergence to suggest that scenario. We encountered this species in three of our four collection localities (all lowland), including the relatively altered forest surrounding Palumeu (Site 5), where they were quite common. We also encountered *Dendrobates tinctorius* (see page 29), which is found only within the GS, but is unique in possessing highly variable, geographically isolated color patterns. Although there is some taxonomic contention whether the different morphs should be treated as unique taxa, genetic data suggests the different color morphs (i.e. populations) are representative of *within* species variation only (Noonan and Gaucher 2006). This taxon was only observed at Sites 1 and 2 (including egg masses and tadpoles at Site 1), although two strikingly different color morphs were found at each locality. At the Grensgebergte Mtns. Site (2), L. Alonso and P. Naskrecki encountered a differently colored morphotype of *D. tinctorius* compared to the lowland coloration found near Site 1, although based solely on a partially out of focus photo voucher we believe it to represent the "common" color morph.

Bufonidae: This hyper-diverse, globally distributed family contains >35 genera with >500 described species, with a number of genera and species endemic to South

America. Although only seven species were encountered on this RAP, they were by far the most abundant amphibians at each of the lowland sites (Sites 1, 4 and 5). The genus *Amazophrynella* was recently separated from *Dendrophryniscus* (Fouquet et al. 2012a), and of the two described and one undescribed species, one—*A. minuta* (see page 29)—has a broad AGR distribution that includes Suriname. Although the common name—tree toads—suggests an arboreal existence, we encountered numerous individuals at Sites 1 and 4 during the day hopping around in the leaf litter. Although we are fairly certain of their taxonomic placement based on morphology, evidence suggests this taxon is actually a species group (Coloma et al. 2010; Fouquet et al. 2012a). Further taxonomic work is required, albeit beyond the scope of this report.

Rhaebo guttatus also has a widespread pan Amazo-Guianan distribution (Azevedo-Ramos et al. 2010). This prodigious species is terrestrial and nocturnal, and was common where it occurred (Sites 1 and 4). As it requires undisturbed primary forest, this species was absent from disturbed, human-modified areas (Site 5), where it was instead replaced by the more ecologically tolerant cane toad, *Rhinella marina* (e.g. observed around human habitation and in a maintained airstrip). Members of the genus *Rhinella* are generally nocturnal, explosive breeders, and representative species were found at all four sampled sites during this RAP survey. *Rhinella marina* has a large native distribution spanning from central South to southern North America, but is considered an invasive species in numerous other countries, notably Australia and the US (Solís et al. 2009). We also encountered at least three other species in this genus, *R. lescurei*, *R. margaritifera* and *R. martyi* . *Rhinella lescurei* and *R. martyi* are described species of the *R. margaritifer* species group (Fouquet et al. 2007), which is broadly distributed in primary rainforest across northern South America (Solís et al. 2009). We encountered a fourth form, *R.* sp., that we believe could be an additional member of the *R. margaritifera* group. Unfortunately, many of the specimens we collected were juveniles, and the named species are quite similar morphologically, so accurate identification has proven difficult. Further work is required.

Centrolenidae: Members of this family are commonly called glass frogs, as their transparent venter makes visible their internal organs. They are nocturnal, colored in shades of neon green and are often found in overhanging vegetation along streams and rivers, although their coloration and behavior make them particularly difficult to locate. Suriname has at least five species of centrolenids (Ouboter and Jairam 2012). We collected just two specimens on this RAP survey, both tentatively identified as *Hyalinobatrachium* cf. *taylori*, which is widespread across the GS. Only one specimen was collected from each of the sites where they were present (Sites 1 and 4), although numerous males were heard calling at both places.

Hylidae: This mega-diverse family is composed of "true tree frogs and their allies" comprising 900+ species in 45+ genera that are distributed mainly in the New World (particularly South and Central America) as well as Australia. They are nocturnal, generally arboreal and display a variety of reproductive modes. There are ~40 species known from Suriname (Ouboter and Jairam 2012), however, just 15 were encountered on this RAP, representative of six genera. We collected six species of *Hypsiboas* treefrogs, including a putatively undescribed taxon, *H.* sp. "chocolate" (see page 23) and four species of *Scinax*, including a putatively undescribed taxon (see page 23) that we believe could represent a Surinamese population of a novel species proposed by Fouquet et al. (2007). The genus *Scinax* is in great need of revision and specimens we collected could represent novel species, but further evidence is required to better understand genetic patterns and species boundaries. These species were not particularly common, but representative hylids were present at each site—although not all species were present at each site. A single specimen of the species *Trachycephalus coriaceus* and two *Dendropsophus* cf. *brevifrons* specimens were recorded (one via cell phone camera photo). The former species has an interesting disjunct distribution with populations present in the eastern GS and southwestern Amazonia, but absent from northeastern Amazonia. Frogs from the genus *Osteocephalus* representing two species (*O. taurinus* and *O. leprieuri*) were some of the most commonly encountered vertebrates at Site 1, yet only a single individual of the latter was found at any of the other sites (Site 4). Lastly, we also recorded the charismatic tiger leg monkey frog species, *Phyllomedusa tomopterna* (see page 29) from Site 1. This species requires pristine forest habitat and is distributed widely across the AGR (La Marca et al. 2004).

Leptodactylidae: Colloquially known as southern frogs, this group is composed of 190 species in 13 genera, all of which are restricted to the New World. This group of frogs is diverse in body size (e.g. 26mm SVL in *Adenomera heyeri* vs. 185mm in *Leptodactylus pentadactylus*) and in ecology (e.g. *Lithodytes lineatus* is often associated with ant nests, whereas *Leptodactylus leptodactyloides* prefers open areas like savannahs and forest edges). We found up to 12 different species (two currently identified to "sp.") from three genera, ~60% of Suriname's total leptodactylid diversity. Members of this family were observed at all four sites, with the greatest diversity of species from Sites 1 and 5 (i.e. Upper Palumeu and Palumeu). Most of the species we encountered are assignable to the genus *Leptodactylus*, although we also collected two forms we cannot yet accurately identify: *Adenomera* sp. and *Leptodactylus* sp.. With the exception of *L. longirostris* and *L. myersi* (which have a more restricted distribution in the GS), the species we encountered are all widely distributed across the AGR—and a few even extend into southern Central America. At site 2, *L. myersi* was the most abundant terrestrial vertebrate observed and it was quite easy to find juvenile frogs living in the moist spaces under nearly every granite exfoliation in the seeps on the exposed inselberg face.

Craugastoridae: This family contains the most speciose genus of vertebrates, *Pristimantis*, with over 400 species

(~5 in Suriname) of direct-developing frogs (i.e. lacks a free-living larval stage). Their deviation from the reproductive strategy norm, liberating them from semi-/permanent water sources, may be one reason for their widespread distribution and successful speciation in a variety of habitats. One described species was encountered (*Pristimantis chiastonotus*), which is distributed throughout the eastern GS. This species is found in leaf litter, generally at low altitudes (<700m asl), and it has been suggested that they are able to cope with some degree of habitat disturbance (Gaucher and Rodrigues 2004). We also collected individuals not immediately identifiable to a named species (which we are calling *Pristimantis* sp.) (see page 23). We found *Pristimantis* species at all four collection localities, including disturbed habitat (i.e. Palumeu) and in the Grensgebergte Mtns. (roughly 750m asl).

Class Amphibia, Order Gymnophiona

Rhinatrematidae: This family of caecilians (a fossorial group of primitive, limbless amphibians) is composed of two species-poor genera endemic to South America. Only one species was encountered on the RAP, *Rhinatrema bivitattum* (see page 28), which appears to be distributed across the northern Guiana Shield, from Guyana to Brazil (Gaucher et al. 2004). It was previously only known from the Brownsberg region of Suriname (Nussbaum and Hoogmoed 1979) and thus our record represents a significant southern range extension. Our local guides collected a single specimen while clearing the area of vegetation for the tent camp at Site 1.

Class Reptilia, Order Squamata

Gekkota: This diverse lizard group has a near worldwide distribution as geckos are found on every continent but Antarctica. Representatives of three families (Sphaerodactylidae, Phyllodactylidae and Gekkonidae) were encountered during this RAP survey. All three families have representative species from the New and Old Worlds, as well as northern and southern hemispheres, although the history of occupation in the New World for sphaerodactylids and phyllodactylids is much older than for gekkonids. Gamble et al. (2011) found evidence to suggest that modern New World sphaerodactylids reached northern South America in the Cretaceous before the break-up of Gondwana (i.e. Gondwanan vicariance), and phyllodactylids arrived shortly thereafter (either via vicariance or dispersal), whereas geckos in the family Gekkonidae reached the New World much later (i.e. within the last 5-10 mil. yrs.) via long-distance, over-water dispersal. We only encountered one species of gekkonid on this RAP, *Hemidactylus mabouia*, which is native to Africa, although it can now be found throughout South and Central America and the Caribbean (via either natural or anthropogenic forces of dispersal). This species was common on the main building of the METS resort in Palumeu. The sphaerodactylid gecko species, *Gonatodes humeralis*, was also fairly common in primary forest and forest edges (Sites 1, 4 and 5) and is distributed widely across northern South America. Although neither was common,

the two other sphaerodactylids encountered, *G. annularis* (see page 28) and the rather diminutive *Pseudogonatodes guianensis*, are restricted to moist microhabitats generally in lowland forest (although the former can be found at ~800m on the Tafelberg; Ouboter pers. comm.). We consider encountering both of these species as a good indicator of forest health. Lastly, we encountered the phyllodactylid gecko *Thecadactylus rapicauda*, a prodigious species that reaches 125mm in snout-vent length. What was once considered one widely distributed taxon found throughout northern South America, Central America and the Lesser Antilles, this species has recently been split into three (*T. solimoensis* is restricted to the western Amazon and *T. oskrobapreinorum* from Sint Maarten). This species is sometimes commensal with man-made structures.

Lacertiformes: This morphologically and ecologically diverse group (sensu Townsend et al. 2004) is distributed throughout the Americas and includes three lizard families: Teiidae, Gymnothalmidae and Amphisbaenidae. Teiid and gymnothalmid lizards were the most commonly encountered reptiles at each site during this RAP. Species of lizards in these families are active hunters and were commonly encountered moving through the leaf litter (e.g. *Alopoglossus, Arthrosaura, Leposoma*), in streams/pools (e.g. *Neusticurus*) (see page 28), under logs (e.g. *Gymnopthalmus*), or moving in the open near tree falls and around camp (e.g. *Ameiva, Kentropyx*). The existence of *Cercosaura argulus* in Suriname was previously based on one record from Palumeu (Hoogmoed 1973) and the distribution map provided by the IUCN website does not confirm its presence in the country (Doan and Avila-Pires 2010). We here provide confirmation of its residence in southern Suriname. Additionally, our tentative taxonomic designation for *Alopoglossus buckleyi* would suggest a significant range extension for this taxon. There are morphological differences separating this taxon from the widespread *A. angulatus* (corroborative genetic data is also being gathered), however, this designation is still tentative and will require more comparative material for us to be confident. A single worm lizard putatively identified as *Amphisbaena* cf. *vanzolinii* (see page 28) was encountered during the second night at Site 4 moving through our recently constructed camp. This taxon—like most amphisbaenian species—is infrequently collected due to its fossorial habits but it appears to be patchily distributed across the western AGR.

Scincomorphs: Skinks are the second most diverse squamate group (behind Gekkota) and are also distributed nearly worldwide, although there are only ~18 species in South America. We encountered a single species, *Mabuya nigropunctata*, at all forest sites—including high elevation—and were commonly observed basking or foraging around open canopy tree falls. *Mabuya nigropunctata* is widespread across Amazonia and can even be found in St. Vincent in the southern Caribbean. Miralles and Carranza (2010) recently published molecular evidence suggesting this taxon might be a complex of three largely allopatric lineages, although they did not suggest new names for the distinct lineages they recovered. We observed numerous individuals at Sites 1, 2 and 4, yet this wily taxon successfully evaded capture so we have no voucher or genetic material with which to compare to Miralles and Carranza (2010).

Serpentes: According to Ávila Pires (2005), Suriname possesses more than 100 snake species from 8 families. Although most of our records were of single individuals, we encountered 19 species from 6 different families. While this may seem like a paltry sum compared to the total, snakes in general are difficult to collect due to their cryptic biology, and some are restricted to specific habitat types/food sources (e.g. mangroves) that were not targeted or near the sampling sites of this study. By comparison, the two previous RAP surveys to Suriname, Lely/Nassau (RAP 43) and Kwamalasamutu (RAP 63) recorded just 6 and 17 snake species, respectively. The family Colubridae comprised the majority of the species we encountered (13 sp.), including the most commonly encountered species (i.e. the Blunthead Tree Snake (*Imantodes cenchoa*) and two species of ground snake (*Atractus* spp.), although still only a small fraction of Suriname's total colubrid diversity (>75 sp.). Colubridae is the largest snake family and includes about two-thirds of all described snake species (>300 genera, >1,900 species). Site 2 (Grensgebergte) provided the only records of Cope's Snail-eater, *Dipsas copei*, and the fer-de-lance, *Bothrops atrox*. There is some confusion regarding the type locality of the former (i.e. Suriname; see Kornacker 1999), although undoubtedly it is quite rare and is represented by only a handful of specimens collected in Suriname. The fer-de-lance on the other hand is often one of the more common snakes encountered in lowland tropical forests; yet we saw only one, and at >700 m. We posit that this is a result of undersampling rather than absence of the species from the lowland sites and if we had utilized pitfall trap arrays with funnel traps on this survey, we are confident we would have encountered more "common" terrestrial species, such as *Bothrops* spp.

Testudines: Although only a few individuals of two species were observed, we argue that presence of turtles and tortoises are positive signs of limited hunting pressure. Due to their ease of capture and convenient storage, humans have been subsistence hunting tortoises for millennia (Thorbjarnarson et al. 2000) and many cultures in South America continue this practice. A single individual of the flat-headed turtle, *Platemys platycephala*, was observed by R. Jairam at Site 1 foraging in partially flooded lowland forest. At least two different individuals of the yellow-footed tortoise, *Chelonoidis denticulata*, were encountered at the high elevation site 2, sitting in vegetation near slowly flowing seeps, while a third individual was also observed foraging in the lowlands between camp and Kasikasima mountain (Site 4). While by no means an extensive survey, the presence of these species is likely correlated to the seemingly pristine quality of the forests.

CONSERVATION RECOMMENDATIONS

Of the species we encountered, only a fraction has been assessed for the IUCN Red List of Threatened Species, and most are listed as Least Concern or Not Evaluated since they are widely distributed across either the greater Guiana Shield or some portion of Amazonia. Due to myriad factors (e.g. weather, collection technique, collector fatigue), it is likely that we failed to collect all representatives of the herpetofaunal community at each site. With repeated surveys, we expect that our species lists would increase. Given the amount of time available at each site, we found a community that appeared speciose and could be harboring a few putatively "new" species (e.g. *Hypsiboas* sp. "chocolate"). This survey also provided collection records that have contributed to geographic range extensions for species not previously known from Suriname (e.g. *Cercosaura argulus* and *Alopoglossus buckleyi*), as well as new records for particularly rarely encountered species (e.g. *Dipsas copei*, *Rhinatrema bivittatum*, *Amphisbaena* cf. *vanzolinii*). Observing little in the way of disturbance or man-made alterations, we suggest that these sites (with the possible exception of the altered forest near Palumeu) are healthy and productive, and are presumably acting as a corridor for gene flow through this region of the Guiana Shield. The presence of species that are rarely seen or were previously unrecorded in Suriname helps to substantiate that there is (or was) an historical connection between this and surrounding areas. Helicopter flights over the area confirm that the forest is widespread and contiguous, which is hopefully contributing to species/genetic admixture between protected areas. Future conservation/landscape genetic work might confirm a connection between the forests surrounding our main study sites and adjacent protected areas, but we nonetheless advocate that maintaining the pristineness of this corridor should be a priority for healthy ecosystem function and to maintain natural gene flow throughout the Guiana Shield.

ACKNOWLEDGEMENTS

We would specifically like to thank Antoine Fouquet and Philippe Kok for assistance in identifying some of the more taxonomically confusing species. We also wish to thank the other members of the RAP team—both scientists and guides/members of the support team—that contributed to this study via opportunistic collections and/or photographic vouchers.

REFERENCES

Ávila Pires, T. C. S. 2005. Reptiles. In: Hollowell, T. and R. P. Reynolds (eds.). Checklist of the terrestrial vertebrates of the Guiana Shield. Bulletin of the Biological Society of Washington, USA. Number 13. Pp. 25–40.

Azevedo-Ramos, C., E. La Marca, M. Hoogmoed and S. Reichle. 2010. *Rhaebo guttatus. In:* IUCN 2012. IUCN Red List of Threatened Species. Version 2012.1. <www.iucnredlist.org>

Brown, J. L. and E. Twomey. 2009. Complicated histories: three new species of poison frogs of the genus Ameerega (Anura: Dendrobatidae) from north-central Peru. Zootaxa .2049:1–38.

Castroviejo-Fisher, S., P. E. Pérez-Peña, J. M. Padial and J. M. Guayasamin. 2012. A second species of the family Allophrynidae (Amphibia: Anura). American Museum Novitates. Number 3739. Pp. 1–17.

Coloma, L. A., S. Ron, R. Reynolds, C. Azevedo-Ramos, F. Castro. 2010. *In:* IUCN 2012. IUCN Red List of Threatened Species. Version 2012.1. <www.iucnredlist.org>

Doan, T. M. and T. C. S. Avila-Pires. 2010. *Cercosaura argulus. In:* IUCN 2012. IUCN Red List of Threatened Species. Version 2012.2. <www.iucnredlist.org>.

Donnelly, M. A., M. H. Chen, and G. G.Watkins. 2004. Sampling amphibians and reptiles in the Iwokrama Forest ecosystem. Proceedings of the Academy of Natural Sciences of Philadelphia. 154:55–69.

Fouquet, A., M. Vences, M. Salducci, A. Meyer, C. Marty, M. Blanc and A. Gilles. 2007. Revealing cryptic diversity using molecular phylogenetics and phylogeography in frogs of the Scinax ruber and Rhinella margaritifera species groups. Molecular Phylogenetics and Evolution. 43:567–582.

Fouquet, A., R. Recoder, M. Texeira Jr., J. Cassimiro, R. C. Amaro, A. Camacho, R. Damasceno, A. C. Carnaval, C. Moritz and M. T. Rodrigues. 2012a. Molecular phylogeny and morphometric analyses reveal deep divergence between Amazonia and Atlantic Forest species of Dendrophryniscus. Molecular Phylogenetics and Evolution. 62:826–838.

Fouquet, A., B. P. Noonan, M. T. Rodrigues, N. Pech, A. Gilles and N. J. Gemmell. 2012b. Multiple Quaternary Refugia in the Eastern Guiana Shield Revealed by Comparative Phylogeography of 12 Frog Species. Systematic Biology. 61:461–489.

Gamble, T., A. M. Bauer, G. R. Colli, E. Greenbaum, T. R. Jackman, L. J. Vitt and A. M. Simons. 2011. Coming to America: multiple origins of New World geckos. Journal of Evolutionary Biology. 24:231–244.

Gardner, T. A., M. A. Ribeiro-Júnior, J. Barlow, T. Cristina, S. Ávila-Pires, M. S. Hoogmoed and C. A. Peres. 2007. The value of primary, secondary, and plantation forests for a neotropical herpetofauna. Conservation Biology. 21:775–787.

Gaucher, P. and M. T. Rodrigues. 2004. *Pristimantis chiastonotus. In:* IUCN 2012. IUCN Red List of Threatened Species. Version 2012.1. <www.iucnredlist.org>

Gaucher, P., R. MacCulloch, M. Wilkinson and M. Wake. 2004. *Rhinatrema bivittatum.* In: IUCN 2012. IUCN

Red List of Threatened Species. Version 2012.1. <www.iucnredlist.org>

Grant, T., D. R. Frost, J. P. Caldwell, R. Gagliardo, C. F. B. Haddad, P. J. R. Kok, D. B. Means, B. P. Noonan, W. E. Schargel and W. C. Wheeler. 2006. Phylogenetic systematics of dart-poison frogs and their relatives (Amphibia: Athesphatanura: Dendrobatidae). Bulletin of the American Museum of Natural History. Number 299:1–262.

Kok, P. 2008. *Allobates granti. In:* IUCN 2012. IUCN Red List of Threatened Species. Version 2012.1. <www.iucnredlist.org>

Kornacker, P. M. 1999. Checklist and key to the snakes of Venezuela. PaKo-Verlag, Rheinbach, Germany. 270 pp.

La Marca, E., C. Azevedo-Ramos, C. L. Barrio Amorós. 2004. *Allophryne ruthveni. In:* IUCN 2012. IUCN Red List of Threatened Species. Version 2012.1. <www.iucnredlist.org>

Marty, C. and P. Gaucher. 2000. Sound guide to The Tailless Amphibians of French Guiana. Centre bioacoustique, Paris.

Miralles, A. and S. Carranza. 2010. Systematics and biogeography of the Neotropical genus *Mabuya*, with special emphasis on the Amazonian skink *Mabuya nigropunctata* (Reptilia, Scincidae). Molecular Phylogenetics and Evolution. 54:857–869.

Nussbaum, R. A. and M. S. Hoogmoed. 1979. Surinam caecilians with notes on *Rhinatrema bivittatum* and the description of a new species of *Microcaecilia* (Amphibia: Gymnophiona). Zoologische Mededelingen. 54:217–235.

Ouboter, P. E., R. Jairam and C. Kasanpawiro. 2011. A rapid assessment of the amphibians and reptiles of the Kwamalasamutu region (Kutari /lower Sipaliwini Rivers), Suriname. *In:* O'Shea, B.J., L.E. Alonso, & T.H. Larsen, (eds.). A Rapid Biological Assessment of the Kwamalasamutu region, Southwestern Suriname. RAP Bulletin of Biological Assessment 63. Conservation International, Arlington, VA. Pp. 124–127.

Ouboter, P. E., R. Jairam. 2012. Amphibians of Suriname. Vol. 1 of Fauna of Suriname. Brill Academic Pub.

Peres, C.A. 2001. Effects of Subsistence Hunting on Vertebrate Community Structure in Amazonian Forests. Conservation Biology. 14:240–253.

Reynolds, R., M. Hoogmoed, R. MacCulloch and P. Gaucher. 2004. *Anomaloglossus baeobatrachus. In:* IUCN 2012. IUCN Red List of Threatened Species. Version 2012.1. <www.iucnredlist.org>

Señaris, J. C. and R. MacCulloch. 2005. Amphibians. In: Hollowell, T. and R. P. Reynolds (eds.). Checklist of the terrestrial vertebrates of the Guiana Shield. Bulletin of the Biological Society of Washington. USA. Number 13. Pp. 25–40.

Solís, F., R. Ibáñez, G. Hammerson, B. Hedges, A. Diesmos, M. Matsui, J.-M. Hero, S. Richards, L. Coloma, S. Ron, E. La Marca, J. Hardy, R. Powell, F. Bolaños,

G. Chaves and P. Ponce. 2009. *Rhinella marina. In:* IUCN 2012. IUCN Red List of Threatened Species. Version 2012.1. <www.iucnredlist.org>

Starace, F. 1998. Guide des Serpents et Amphisbenes de Guyane. Ibis Rouge, Guadeloupe/Guyane. 449 pp.

Stuart, S.N., J.S. Chanson, N.A. Cox, B.E. Young, A.S.L. Rodrigues, D.L. Fischman and R.W. Waller. 2004. Status and trends of amphibian declines and extinctions worldwide. Science. 5702:1783–1786.

Thorbjarnarson, J., C.J. Lagueux, D. Bolze, M.W. Klemens and A.B. Meylan. 2000. Human use of turtles. In: M.W. Klemens (ed.). Turtle Conservation. Smithsonian Institution, Washington, DC, USA. Pp. 33–84.

Townsend, T. M., A. Larson, E. Louis and J. R. Macey. 2004. Molecular Phylogenetics of Squamata: The Position of Snakes, Amphisbaenians, and Dibamids, and the Root of the Squamate Tree. Systematic Biology. 53:735–757.

Watling, J. I. and L. F. Ngadino. 2007. A preliminary survey of amphibians and reptiles on the Nassau and Lely plateaus Eastern Suriname. *In:* Alonso, L.E. and J.H. Mol (eds.). A rapid biological assessment of the Lely and Nassau plateaus, Suriname (with additional information on the Brownsberg Plateau). RAP Bulletin of Biological Assessment 43. Conservation International, Arlington, VA, USA. Pp. 119–125.

Appendix 9.1. Amphibians and reptiles recorded during the current RAP study. A qualitative assessment of species abundance for each site where the occurred is included (VC: very common, >10 individuals observed; C: common, 1>x >10 individuals observed; UC: uncommon, only 1 observed/captured), as well as general geographic distribution (W: widespread; GS: Guiana Shield; AGR: Amazo-Guianan Subregion; E: exotic), IUCN threat status (LC: least concern; NE: not evaluated; DD: data deficient) and type of microhabitat in which the species was recorded. The last two columns indicate whether the same (Y or N) species were recorded by the two most recent RAP surveys in Suriname. In some cases, congeners (N*; i.e. a close taxonomic relative) were recorded instead of the species we documented in the present study.

Taxon	cf.?	Per Locality Qualitative Records				Distribution	IUCN Threat Status	Microhabitat Type	RAP 43	RAP 63
		Upper Palumeu	Grensgebergte	Kasikasima	Palumeu					
AMPHIBIA (47 species)										
AMPHIBIA-ANURA (46 species)										
Allophrynidae										
Allophryne ruthveni		C		UC		GS	LC	lowland forest	N	Y
Aromobatidae										
Allobates femoralis		VC		VC	UC	AGR	LC	lowland forest	Y	Y
Allobates granti	cf.	UC				GS	LC	forest?	N	Y
Anomaloglossus baeobatrachus	cf.			C		GS	DD	lowland forest	N	Y
Anomaloglossus sp.		UC?		VC		??	??	lowland forest	-	-
Bufonidae										
Amazophrynella minuta		VC		VC		AGR	LC	lowland forest, forest stream	N	N
Rhaebo guttata		C		C		AGR	LC	forest (high/lowland)	Y	Y
Rhinella lescurei	cf.	C		C		GS	LC	lowland forest	N	Y
Rhinella margaritifer	cf.	VC		VC		W	LC	lowland forest	Y	N
Rhinella marina				C	VC	W	LC	lowland forest	Y	Y
Rhinella martyi	cf.	C	C			GS	LC	lowland forest	N	Y
Rhinella sp.		C		C		??	??	lowland forest	-	-
Centrolenidae										
Hyalinobatrachium taylori	cf.	C		C		GS	LC	forest stream side	Y	Y
Dendrobatidae										
Ameerega trivittata		VC		VC	C	AGR	LC	forest	Y	Y
Dendrobates tinctorius		C		C		GS	LC	lowland forest	N	Y

table continued on next page

Appendix 9.1. continued

table continued on next page

Taxon	cf.?	Per Locality Qualitative Records				Distribution	IUCN Threat Status	Microhabitat Type	RAP 43	RAP 63
		Upper Palumeu	Grensgebergte	Kasikasima	Palumeu					
Hylidae										
Dendropsophus brevifrons	cf.			UC		AGR (patchy)	LC	lowland forest	N*	N*
Hypsiboas boans			UC	UC		AGR	LC	lowland forest	Y	Y
Hypsiboas calcaratus		C				AGR	LC	lowland forest	N	Y
Hypsiboas fasciatus		C		UC		AGR	LC	lowland forest	N	Y
Hypsiboas geographicus				UC		AGR	LC	lowland forest	N	Y
Hypsiboas ornatissimus					UC	GS	LC	lowland forest	N	N
Hypsiboas sp.				UC		??	??	lowland forest	-	-
Osteocephalus leprieuri		C		UC		AGR	LC	lowland forest	N	Y
Osteocephalus taurinus		C				AGR	LC	lowland forest	Y	Y
Phyllomedusa tomopterna		C				AGR	LC	lowland forest	N*	N*
Scinax cruentommus		C				AGR	LC	lowland forest	N*	N*
Scinax ruber	cf				UC	W	LC	lowland forest	N*	N*
Scinax sp. "hybrid"		C				??	??	lowland forest	N*	N*
Scinax x-signatus	cf.				UC	W	LC	lowland forest	N*	N*
Trachycephalus coriaceus	cf.	UC				AGR (patchy)	LC	forest stream side	N	N*
Leptodactylidae										
Adenomera heyeri				C		GS	LC	lowland forest	N*	Y
Adenomera sp.			C			??	??	lowland forest	-	-
Leptodactylus bolivianus					VC	AGR	LC	open grassland, human habitation	N	Y
Leptodactylus knudseni					C	AGR	LC	open grassland, human habitation	Y	Y
Leptodactylus leptodactyloides	cf.	C		C	C	AGR	LC	lowland forest	Y	N
Leptodactylus lineatus					C	W	LC	modified forest, human habitation	N	N
Leptodactylus longirostris		UC				GS	LC	lowland forest	Y	N
Leptodactylus myersi			VC	UC		GS	LC	granite outcrop, human hab.	N	Y
Leptodactylus mystaceus		UC			UC	AGR	LC	lowland forest, human hab.	Y	Y
Leptodactylus pentadactylus		UC				AGR	LC	lowland forest	Y	Y
Leptodactylus rhodomystax		UC				AGR	LC	lowland forest, human hab.	N	Y
Leptodactylus sp.	cf.	UC			UC	??	??	lowland forest, human hab.	-	-
Leptodactylus stenodema	cf.			UC		AGR	LC	lowland forest	Y	N

Appendix 9.1. continued

Taxon	cf.?	Per Locality Qualitative Records				Distribution	IUCN Threat Status	Microhabitat Type	RAP 43	RAP 63
		Upper Palumeu	Grensgebergte	Kasikasima	Palumeu					
Strabomantidae										
Pristimantis chiastonotus	cf.	C	C	C		GS	LC	lowland forest, granitic forest	Y	Y
Pristimantis sp.		C	UC	C		??	??	varied	-	-
Pristimantis zeuctotylus					UC			forest edge	Y	N
AMPHIBIA-GYMNOPHIONA (1 sp.)										
Rhinatrematidae										
Rhinatrema bivitattum		UC				GS	LC	lowland forest	N	N*
REPTILIA (42 sp.)										
SQUAMATA-GEKKOTA (5 sp.)										
Gekkonidae										
Hemidactylus mabouia	cf.				VC	E	LC	human habitation	N	N
Sphaerodactylidae										
Gonatodes annularis		C		C		W	NE	lowland forest	Y	Y
Gonatodes humeralis		C		C	C	W	NE	forest gaps/edges	Y	Y
Pseudogonatodes guianensis		UC				AGR	NE	lowland forest	N	N*
Phyllodactylidae										
Thecadactylus rapicauda		C				W	NE	lowland forest	N	Y
SQUAMATA-LACERTIFORMES (12 sp.)										
Amphisbaenidae										
Amphisbaena vanzolinii	cf.			UC		AGR (patchy)	DD	lowland forest	N*	N*
Gymnophthalmidae										
Alopoglossus angulatus		UC				AGR	LC	lowland forest	N	N
Alopoglossus buckleyi	cf.		UC			??	NE	granitic forest	N	N
Arthrosaura kockii				UC		GS	LC	lowland forest	Y	Y
Arthrosaura reticulata		C				W	NE	lowland forest	N	N
Gymnophthalmus underwoodi					UC	GS*	LC	modified forest edge	N	Y
Leposoma guianense	cf.	C		C	UC	GS	NE	1o/2o/granitic forest	Y	Y
Neusticurus bicarinatus		UC				AGR	NE	lowland forest	N*	Y
Cercosaura argulus	cf.	C				AGR	NE	lowland forest	N*	N
Tretioscincus agilis		UC		UC		AGR	NE	lowland forest	N	N

table continued on next page

Appendix 9.1. continued

table continued on next page

Taxon	cf.?	Per Locality Qualitative Records				Distribution	IUCN Threat Status	Microhabitat Type	RAP 43	RAP 63
		Upper Palumeu	Grensgebergte	Kasikasima	Palumeu					
Teiidae										
Ameiva ameiva		VC	VC	VC	VC	W	NE	varied	Y	N
Kentropyx calcarata		VC	VC	VC	VC	AGR	NE	forest gaps in 1o forest	Y	Y
SQUAMATA–IGUANIA (3 sp.)										
Polychrotidae										
Norops nitens	cf.	C		C		AGR	NE	lowland forest	N*	Y
Tropiduridae										
Plica (=Tropidurus) umbra				C		AGR	NE	lowland forest	N*	Y
Uranoscodon superciliosus	cf.	C				AGR	NE	lowland forest	N	N
SQUAMATA–SCINCOIDEA (1 sp.)										
Scincidae										
Mabuya nigropunctata		C	C	C		W	NE	lowland/granitic forest; tree falls	Y	Y
SQUAMATA–SERPENTES (19 sp.)										
Aniliidae										
Anilius scytale				UC		W	NE	lowland forest	N	Y
Boiidae										
Corallus caninus				UC		AGR	NE	river edge	N	N*
Epicrates cenchria				UC		W	NE	lowland forest	N	N
Colubridae										
Atractus flammigerus	cf.			UC		AGR	NE	lowland forest	N	Y
Atractus torquatus	cf.	C				AGR	NE	lowland forest	N	Y
Chironius exoletus		UC				W	NE	lowland forest	N*	N
Chironius fuscus		UC				W	NE	lowland forest	N*	N
Dipsas copei			UC			GS	NE	granitic forest	N*	N*
Erythrolamprus aesculapi		UC				AGR	NE	lowland forest	N	Y
Helicops angulatus	cf.	UC				AGR	NE	lowland forest stream	N	Y
Imantodes cenchoa		C		C		W	NE	lowland forest	N*	Y
Oxybelis argenteus		UC				W	NE	lowland forest	N	N
Oxyrhopus formosus	cf.			UC		AGR	NE	lowland forest	Y	Y
Pseudoboa sp.				UC		??	??	lowland forest	-	-
Psustes poecilonotus				UC		W	LC	lowland forest stream	N	N
Tripanurgos (Siphlophus) compressus		UC				W	NE	lowland forest	N	N*

Appendix 9.1. continued

| Taxon | cf.? | Per Locality Qualitative Records | | | | Distribution | IUCN Threat Status | Microhabitat Type | RAP 43 | RAP 63 |
		Upper Palumeu	Grensgebergte	Kasikasima	Palumeu					
Elapidae										
Micrurus psyches		UC				GS	NE	lowland forest	N	N
Typhlopidae										
Epictia (Leptotyphlops) tenella				UC		W	NE	lowland forest clearing (i.e. camp)	N	N*
Viperidae										
Bothrops atrox			UC			AGR	NE	high altitude grassland	Y	Y
TESTUDINES (2 sp.)										
Chelidae										
Platemys platycephala		UC				AGR	NE	lowland, flooded forest	Y	Y
Testudinidae										
Chelonoidis denticulata			C	UC		W	LC	high altitude grassland, lowland forest	N	Y
CROCODYLIA (1 sp.)										
Alligatoridae										
Paleosuchus sp.						AGR			Y	Y

Chapter 10

A rapid assessment of the avifauna of the upper Palumeu watershed, Southeastern Suriname

Brian J. O'Shea and Serano Ramcharan

SUMMARY

We present the results of ornithology surveys carried out during the SE Suriname RAP expedition, 8–29 March 2012. Birds were surveyed using line transect counts and casual observation in lowland forest around the Juuru and Kasikasima camps. A limited survey using mist nets was undertaken in high-elevation (800 m) savanna forest and scrub in the Grensgebergte. Our list of 313 species includes all birds seen or heard at the two RAP camps, the high-elevation satellite camp, the village of Palumeu, and during excursions along the Palumeu River. We recorded fourteen species listed as Vulnerable or Near-Threatened on the IUCN Red List, and consider another seven species as likely to occur in the region. Our records of several species represent range extensions within Suriname and the Guiana Shield. Whereas the lowland forest avifauna was broadly similar at the different localities, 32% of species were only observed at one of the four survey sites. The abundance of parrots and cracids was particularly noteworthy, especially compared to the more populated Kwamalasamutu region that we surveyed in 2010. The high-elevation savanna forest harbored several species not known to occur in the adjacent lowlands, and therefore had the most unique species assemblage of any site. Our results indicate that the lowland forest of SE Suriname probably contains the vast majority of bird species known to occur in the country's interior, including many species of high conservation value, arguing strongly for protection of the region's forests. We recommend further surveys of high-elevation sites in the Grensgebergte and other mountain ranges in southern Suriname, to better determine the range limits of species restricted to high-elevation forests.

INTRODUCTION

Birds are an important component of the ecology of tropical forests—they are major predators of arthropods and small vertebrates, and function as the primary dispersers of many tree species. As in many other taxonomic groups, species diversity is generally high at forested lowland sites in the tropics. Because most birds are diurnal and many can be identified by sound alone, they can be surveyed relatively quickly. In addition, birds are generally ubiquitous, and many species are colorful—traits that render them appealing to nature-oriented tourists, whose revenue can provide an important contribution to local economies. Birds are therefore an ideal study group for rapid biodiversity surveys.

The avifauna of Suriname is well known (Ottema et al. 2009) though new records continue to accumulate as more interior localities are inventoried, particularly in the southern half of the country (O'Shea 2005, Mittermeier et al. 2010, Zyskowski et al. 2011, O'Shea and Ramcharan 2011). Much of the interior of Suriname is covered by unbroken tropical moist forest with few human settlements. Accordingly, the avifauna is diverse, and the country's forests support healthy populations of species that are of global conservation concern, such as large raptors, guans and curassows (Cracidae, hereafter "cracids"), and parrots. Many lowland species, though not endemic to Suriname itself, are nonetheless restricted to the Guiana Shield; the forests of Suriname are vital for the persistence of these species.

In addition to widespread lowland species, the mountains and plateaus of interior Suriname support a suite of species with highland affinities, tending not to occur below approximately 400 m. The distribution of these species within Suriname is not well understood, primarily because many highland areas in central and southern Suriname are extremely difficult to access. Highland species are particularly interesting because they persist as more-or-less isolated populations across the Guiana Shield, yet none of these species have been sampled adequately to determine the extent of morphological and genetic variation among their populations. Because little is known of their ecology and distribution, their vulnerability to climate change is difficult to assess, though this is an urgent priority in conservation planning for these species in the Guianan highlands.

The goal of this survey was to develop baseline data on the avifauna of selected localities in SE Suriname, with a particular focus on high-elevation forests in the Grensgebergte. This RAP was the second in a series of expeditions documenting the flora and fauna of southern Suriname; therefore, a secondary goal was to compare the avifauna with that of the Kwamalasamutu region surveyed in 2010 (O'Shea and Ramcharan 2011). We also sought to confirm the presence of rare and endemic species, expand the known ranges of species with limited distributions in the Guianas, and assess the population status of several species important to the Trio and Wayana people, either as food or as flagship species for ecotourism.

STUDY SITES AND METHODS

Birds were surveyed around four principal sites: the Juuru and Kasikasima camps, the Grensgebergte satellite camp, and the village of Palumeu. A map and coordinates of these localities are presented elsewhere in this report. The sites were surveyed on the following dates:

Juuru: 9–11, 13–14, 17–18 March. The Juuru camp was situated on the left bank of the upper Palumeu River, at the base of a steep hillside. From this camp we were able to survey terra firme forest on steep slopes at 270–420 m. Across the river, the forest was tall and seasonally inundated on level to gently sloping terrain, with many *Euterpe* palm swamps and a small open area dominated by *Guadua* bamboo.

Kasikasima: 20–28 March. The Kasikasima camp was situated along the left bank of the Palumeu River, approximately 61 km northeast of the Juuru camp and three km east of Kasikasima itself. The habitat at this site was primarily tall terra firme forest on rolling terrain, with occasional steep-sided creek valleys, some of them dominated by *Euterpe* palms. Around the base of Kasikasima, the forest was drier and shorter in stature.

Grensgebergte: 12, 14–16 March. The Grensgebergte satellite camp was at the summit of an unnamed mountain at an elevation of 790 m, approximately 16 km northwest of the Juuru camp; from here we were able to sample a limited area of forest habitat up to approximately 820 m. The habitat at this site was "savanna" forest consisting of densely spaced trees less than 30 cm dbh; this forest grew wherever there was a substantial soil layer, primarily along the ridge that formed the top of the mountain, and ranged from 10–30 m in height. Vegetation on rocky areas and steep slopes was dominated by ground bromeliads and grasses, with occasional small trees and substantial areas of open rock.

Palumeu: 8–9, 28–29 March. Aside from a brief early-morning boat trip near the village on 9 March, most survey effort around Palumeu was concentrated along the airstrip and around the tourist facilities maintained by METS.

At the Juuru and Kasikasima camps, 200-m line transects were established to gather quantitative data on bird species composition and abundance. Transects were spaced as far apart as logistics allowed, but were never closer than 200 m from one another. Three transects were sampled at the Juuru camp and five at the Kasikasima camp. Each transect was sampled only once. Starting at first light (0630–0635), one observer walked slowly for 30 minutes along the complete length of a transect, counting all birds seen or heard. Transects were not conducted in rainy conditions, which precluded the establishment of any transects at the Grensgebergte camp, where rain at dawn was a daily event. Transect data were analyzed using EstimateS Version 8.2 (Colwell 2006); because few species were represented by more than 3 individuals on any transect, we calculated the Chao 2 estimator (Chao 1987), which estimates species richness based on incidence rather than abundance. We also applied Chao's Sørensen Similarity Index (Chao et al. 2005) to the transect data to assess community similarity among camps. To place the avifauna of the study areas in a broader geographic context, we included transect data gathered during the 2010 Kwamalasamutu RAP survey (O'Shea and Ramcharan 2011).

Casual observations of the avifauna were made by walking along trails to locate and identify birds, with an emphasis on concentrated food sources (e.g., fruiting and flowering trees), mixed-species foraging flocks, and vantage points where large areas of canopy or sky could be viewed. Birds were also surveyed during the course of boat excursions along the Palumeu River on two afternoons. A list of observed species was compiled at the end of each day.

At the Grensgebergte camp, birds were captured in 12×2-m ground-level mist nets set in low savanna forest at 820 m. Six nets were deployed for two days and opened primarily during the afternoon hours. Nets were kept closed during rains and at night. Birds captured in the nets were photographed and released.

Birds were documented using Marantz PMD-661 and PMD-660 digital sound recorders. To record the dawn soundscape, a stereo microphone pair was operated remotely for 2–3 hours on several mornings near the Kasikasima camp. Recordings will be deposited at the Macaulay Library at the Cornell Lab of Ornithology in Ithaca, New York, USA.

RESULTS

We observed a total of 313 species during the RAP expedition; these are listed in Appendix 10.1. At each of the main RAP camps (Juuru and Kasikasima), we found a highly diverse avifauna consisting of species typical of both terra firme and seasonally inundated forest. Our survey efforts at the Juuru camp were hindered by persistent rain, which forced the cancellation of two transect surveys and flooded much of the forest surrounding the camp, rendering it inaccessible. As a result, we observed fewer species at the Juuru site (196) than at the Kasikasima site (233), though our transect data suggest that both localities harbor comparable bird diversity (Table 10.1). Chao 2 estimates of species richness from transect data were not significantly higher for the SE Suriname RAP camps than for the Kwamalasamutu RAP camps (t = 1.70, p = 0.19). Across all sites, the Chao 2 estimator tended to underestimate diversity (Table 10.1);

Table 10.1. Chao 2 estimates of species richness for the SE Suriname and Kwamalasamutu RAP surveys based on transect data.

Site	Number of transects	Species observed on transects	Chao 2 Mean (Upper 95% CI)	Total species observed at site
Kasikasima	5	80	141.63 (215.66)	233
Juuru	3	71	156.75 (255.34)	196
Werehpai	3	79	147.44 (220.01)	221
Sipaliwini	5	88	134.69 (190.48)	250
Kutari	5	89	143.56 (201.35)	216

Table 10.2. Values of Chao's Sørensen similarity index based on pairwise comparison of transect data from the Kwamalasamutu and SE Suriname RAP camps. Higher values indicate greater similarity in species composition. Sample sizes (number of transects per camp) are given in parentheses.

	Kasikasima	Juuru	Werehpai	Sipaliwini
Juuru	.789 (3)	-		
Werehpai	.765 (3)	.818 (3)	-	
Sipaliwini	.839 (5)	.630 (3)	.916 (3)	-
Kutari	.944 (5)	.964 (3)	1.000 (3)	.910 (5)

this is not surprising, given that fewer than half of species recorded at each site were detected on transect surveys (range 34.3–41.2%; mean = 36.5%).

The avifauna around the Juuru and Kasikasima camps contained many of the same species recorded on the Kwamalasamutu RAP survey (O'Shea and Ramcharan 2011). Chao's Sørensen index (Chao et al. 2005) predicted substantial similarity in species composition among our survey localities in southern Suriname, irrespective of distance or season (Table 10.2); this is corroborated by our own observations, with the twin caveats that certain species are more likely than others to be recorded on transect surveys by virtue of behavior alone, and that our transect methodology is biased toward species that tend to vocalize at dawn. The high similarity index values between Kwamalasamutu and SE Suriname RAP localities suggest that seasonal effects on species composition were negligible, although seasonal changes in vocal activity may have contributed to dramatic differences in the perceived abundances of many species.

Despite the superficial similarity in species composition among sites suggested by the transect data, there were salient differences as well. Of the 313 species observed on the RAP, 101 (32%) were observed at only one of the four sites. Juuru harbored 24 site-unique species, which is noteworthy considering the inclement weather and our limited mobility at this site relative to Kasikasima. Forty species were observed only at Kasikasima. The majority of the 22 species observed only in the vicinity of Palumeu are widespread species found commonly in disturbed habitats in the Guianas and much of Amazonia. On the other hand, we observed 15 species only at the Grensgebergte satellite camp, and most of these are

indeed restricted to high elevations where they occur in the Guianas (Table 10.3). The proportion of site-unique species, nearly one-third of the total, suggests important underlying environmental differences among these sites that may not be apparent from casual observation.

Overall, we found high heterogeneity in bird species richness and abundance between the Juuru and Kasikasima camps, and between the SE Suriname RAP sites and those surveyed during the Kwamalasamutu RAP in 2010. Differences in species richness among localities were not apparent from the transect data, and were largely due to our sporadic observations of rare and inconspicuous species which comprise the majority of bird species in tropical lowland forest. Differences among sites were also influenced by the patchy nature of certain habitats (e.g. inselbergs, bamboo) and the bird species restricted to those habitats, and, to a lesser degree, regional variation in the abundance of widespread bird species coupled with seasonal changes in vocal behavior. These factors could have rendered certain species considerably easier to detect at some sites than others.

RARE AND THREATENED SPECIES

The recent revision of the IUCN Red List (IUCN 2012) includes 38 species known to occur in the Guiana Shield region; this represents a substantial increase from previous versions, and highlights the region's high level of endemism and the vulnerability of "game" birds to human encroachment on the region's forests. Twenty-nine Near-Threatened (NT) and Vulnerable (VU) species occur in Suriname; we recorded 14 of these on the RAP survey (see Appendix 10.1). An additional seven species that we did not observe during the RAP survey likely occur in the upper Palumeu watershed during all or part of the year: Agami Heron (*Agamia agami*; VU), Gray-bellied Hawk (*Accipiter poliogaster*; NT), Harpy Eagle (*Harpia harpyja*; NT), Crested Eagle (*Morphnus guianensis*; NT), Ornate Hawk-Eagle (*Spizaetus ornatus*; NT), Blue-cheeked Parrot (*Amazona dufresniana*; NT), and Olive-sided Flycatcher (*Contopus cooperi*; NT).

Both cracids and parrots were noted frequently on the SE Suriname RAP survey; by contrast, we observed relatively few at sites within a day's travel by boat from the village of Kwamalasamutu. This is perhaps the most significant difference in the avifaunas of the two survey areas. We observed

Table 10.3. Bird species observed in low savanna forest and scrub between 790–820 m at the Grensgebergte satellite camp. Species highlighted in bold were not observed at any of the other survey sites. Birds seen only in tall forest at this site are excluded; see Appendix 10.1 for complete species list.

Crypturellus soui	Little Tinamou
Ortalis motmot	Variable Chachalaca
Ictinia plumbea	Plumbeous Kite
Anurolimnas viridis	Russet-crowned Crake
Piaya cayana	Squirrel Cuckoo
Nyctidromus albicollis	Common Pauraque
Phaethornis augusti	**Sooty-capped Hermit**
Campylopterus largipennis	Gray-breasted Sabrewing
Thalurania furcata	Fork-tailed Woodnymph
Amazilia viridigaster	**Green-bellied Hummingbird**
Ramphastos tucanus	White-throated Toucan
Picumnus exilis	Golden-spangled Piculet
Ibycter americanus	Red-throated Caracara
Cercomacra tyrannina	Dusky Antbird
Percnostola rufifrons	Black-headed Antbird
Xiphorhynchus pardalotus	Chestnut-rumped Woodcreeper
Synallaxis macconnelli	McConnell's Spinetail
Mionectes macconnelli	McConnell's Flycatcher
Todirostrum cinereum	Common Tody-Flycatcher
Contopus virens/sordidulus **sp.**	**Eastern/Western Wood-Pewee**
Knipolegus poecilurus	**Rufous-tailed Tyrant**
Myiozetetes cayanensis	Rusty-margined Flycatcher
Megarynchus pitangua	Boat-billed Flycatcher
Tyrannus melancholicus	Tropical Kingbird
Myiarchus ferox	Short-crested Flycatcher
Corapipo gutturalis	White-throated Manakin
Lepidothrix serena	White-fronted Manakin
Cyclarhis gujanensis	Rufous-browed Peppershrike
Hylophilus sclateri	**Tepui Greenlet**
Tachyphonus phoeniceus	**Red-shouldered Tanager**
Ramphocelus carbo	Silver-beaked Tanager
Thraupis episcopus	Blue-gray Tanager
Tangara gyrola	**Bay-headed Tanager**
Cyanerpes cyaneus	Red-legged Honeycreeper
Oryzoborus angolensis	Chestnut-bellied Seed-Finch
Coereba flaveola	Bananaquit
Zonotrichia capensis	**Rufous-collared Sparrow**
Piranga flava	**Hepatic Tanager**
Euphonia cyanocephala	**Golden-rumped Euphonia**

15 species of parrots on the SE Suriname RAP—fourteen of which were recorded from the Kasikasima site alone—and although the diversity of parrots in the Kwamalasamutu region was comparable, their abundance was markedly lower during our survey in 2010. The abundance of larger parrot species (e.g., *Amazona, Ara*) in the more remote regions of SE Suriname is noteworthy, and underscores the importance of these forests in maintaining regional populations of species vulnerable to human persecution.

Large birds of prey were comparatively scarce during the survey, a fact we attribute primarily to the frequent rainy and overcast conditions, which limit these birds' tendency to soar and thus render them more difficult to detect. However, we are confident that large raptors occur throughout the region, as many of them are well known to the Trio and Wayana people.

NOTEWORTHY OBSERVATIONS

Rusty Tinamou (*Crypturellus brevirostris*). This species was first recorded in Suriname during the 2010 Kwamalasamutu RAP survey (O'Shea and Ramcharan 2011). We again heard this species' distinctive call on one occasion at the Kasikasima site, along the trail to Kasikasima from the METS tourist camp. The bird vocalized only once and we were unable to record it. The species therefore remains undocumented for Suriname, although our records suggest that it is a low-density resident of lowland forests in the southern part of the country.

White-tailed Hawk (*Geranoaetus albicaudatus*). On 20 March, while scanning from the 500-m vantage point at the end of the METS trail to Kasikasima, we observed a subadult White-tailed Hawk soaring over the top of the Kasikasima massif. After moving a short distance to the east, the bird returned and perched on a small tree at the edge of the summit. White-tailed Hawk has a wide range in South America, but is typically found in open forest and savanna habitats rather than extensive regions of lowland humid forest. Although White-tailed Hawk is unlikely to occur regularly in southern Suriname away from the Sipaliwini savanna, it seems that the species may occasionally use the patchy scrub habitats that occur on inselbergs across the region.

Violaceous Quail-Dove (*Geotrygon violacea*). This enigmatic species is apparently rare in Suriname (O'Shea 2005) and its habitat requirements are unknown; at most localities, it is far outnumbered or replaced entirely by Ruddy Quail-Dove (*G. montana*). We flushed several quail-doves that appeared to be this species along trails at the Kasikasima site, but we could not get visual confirmation of their identity. However, BJO recorded one singing at length shortly after sunrise on 26 March. Kasikasima is therefore one of the few localities in Suriname where this species is known to occur.

White-winged Potoo (*Nyctibius leucopterus*). The first records for Suriname of this poorly known species have been relatively recent (Ottema et al. 2009) and it is presumed to be a rare resident of lowland forest in the Guianas. We were therefore surprised to hear this species on at least three occasions at the Kasikasima site, where it was recorded singing at dawn on 22 March, and we also heard one from the Grensgebergte satellite camp at sunset on 14 March. This latter record (at 800 m) may represent an altitudinal range extension for the species.

Sooty-capped Hermit (*Phaethornis augusti*). This was the common hermit in savanna forest and open scrub at the Grensgebergte satellite camp (see page 30); we did not find it in tall forest or at lower elevations. In Suriname, the species was previously known only from the Sipaliwini savanna (O'Shea 2005; Mittermeier et al. 2010). Our records represent a small range extension.

Orange-breasted Falcon (*Falco deiroleucus*). On 24 March, we observed an Orange-breasted Falcon as it cruised past our vantage point at 500 m at the end of the METS tourist trail to Kasikasima. This bird was easily separable from the similar Bat Falcon (*F. rufigularis*) by both size and voice as it flew past us at close range. Orange-breasted Falcon has a large geographic range but is very locally distributed, often occurring near rock outcrops like Kasikasima, where we suspect they may breed. If confirmed, Kasikasima would be one of the only known breeding localities for this species in Suriname.

Chestnut-fronted Macaw (*Ara severa*). We found Chestnut-fronted Macaw to be fairly common around the Juuru camp but absent from Kasikasima. This is notable considering that macaws are easy to detect, and indicates a possible distributional limit between the two localities, at least during the rainy season. This species appears to be patchily distributed in the Guiana Shield, as it is generally absent from forests of central Suriname, and also from large areas in northern Pará, Brazil (Aleixo et al. 2011).

Wood-Pewee sp. (*Contopus sordidulus/virens* sp.). On 22 March, we observed a wood-pewee sallying from a dead tree at 800 m near the Grensgebergte satellite camp. It was recognized immediately as a boreal migrant, either Western (*C. sordidulus*) or Eastern Wood-Pewee (*C. virens*). Although these two species can be separated by voice, their plumages are very similar. Plumage characters were more suggestive of *C. sordidulus* than *C. virens*, but we resisted the temptation to assign this bird to either species, as it was silent and lighting conditions were suboptimal. Records of migrant and wintering wood-pewees in the Guiana Shield are relatively few; *C. virens* has recently been reported from Pará, Brazil (Aleixo et al. 2011), whereas only *C. sordidulus* is known from French Guiana (Comité d'Homologation de Guyane 2012). Either species would be new for Suriname.

Rufous-tailed Tyrant (*Knipolegus poecilurus*). During the reconnaissance trip to the vicinity of the Grensgebergte satellite camp on 12 March, BJO observed a Rufous-tailed Tyrant in a patch of low savanna forest at 820 m as it foraged in the forest understory, 3–5 m above the ground, for several minutes. We failed to relocate the bird during our stay on

the mountain from 14–16 March. This species was recently found to occur in the Wilhelmina Mountains of Suriname by K. Zyskowski and colleagues (A. Spaans, *pers. comm.*); the closest previous records are from the tepui region, hundreds of kilometers to the west. These records extend the known range of the species into Suriname, and our observation suggests that it should occur in suitable habitat on higher mountains elsewhere in the country.

Guianan Cock-of-the-rock (*Rupicola rupicola*). We found the Cock-of-the-rock in tall forest at the Juuru, Grensgebergte, and Kasikasima camps; based on our own observations and those reported to us, they seem to be common in this region of Suriname. This spectacular bird is a flagship species for ecotourism.

Tepui Greenlet (*Hylophilus sclateri*). We found Tepui Greenlet to be fairly common in both low and tall forest at ~800 m around the Grensgebergte satellite camp. Although the center of its distribution is in the tepui region far to the west, this species is distributed fairly widely above 500 m in the Guiana Shield, occurring as far east as Tafelberg in central Suriname (Zyskowski et al. 2011) and as far south as the Acarai mountains along the Guyana/Brazil border (O'Shea 2008). Our records from the Grensgebergte represent a further range extension to the southeast.

CONSERVATION RECOMMENDATIONS

The forests of SE Suriname harbor a rich avifauna that most likely includes the vast majority of the ~500 species known to occur in the country's interior. Although there were many similarities in species composition between our two lowland survey sites, and between SE Suriname and the Kwamalasamutu region, there were important differences as well. Perhaps most significant is the difference in abundance of parrots; SE Suriname is clearly a stronghold for parrots, particularly macaws, due primarily to the lack of human settlements and infrastructure that would make these birds more accessible to hunters and trappers. Likewise, the region's remoteness is important for the persistence of cracid populations, which tend to be reduced quickly by hunting, as has occurred in the Kwamalasamutu region. The Cock-of-the-rock, which we found to be common in the region's forests, is a huge draw for nature-oriented tourists, and we recommend that tour operators continue to highlight this species for visitors.

There is some evidence that the avifauna in the southern parts of the Guianas differs from sites further north, a pattern that has also been noted for trees. Virtually all recent records of new bird species for both Suriname and Guyana have come from the southern regions of those countries, and many of those species have yet to be found farther north, strongly suggesting that their distributions are limited by subtle environmental differences between northern and southern forests. This pattern, coupled with the remoteness of SE Suriname and the healthy populations of parrots and cracids found there, argue strongly for protection of this region.

Future plans for mining and hydropower projects in the interior of Suriname pose potential threats to the region's spectacular bird diversity. Infrastructure expansion and large-scale extraction of the region's resources would lead to increased hunting, conversion of forest into open habitats with much less biodiversity, and degradation of remaining forest habitat. One of the greatest threats to the region at present is the continuing influx of small-scale gold miners, whose activities inflict tremendous damage on aquatic ecosystems and cause local depletion of fish and game animal populations. We recommend that gold miners be vigorously excluded from the sensitive headwaters of the Palumeu and other rivers originating in the highlands along Suriname's southern border, as part of a broader effort to protect habitats at these rivers' sources.

The conservation value of the highland forests of the Grensgebergte warrants further investigation. Several of our most interesting records came from the Grensgebergte camp, where we spent the least amount of time, and had the most limited mobility and the worst survey conditions. The highland forest, though relatively limited in extent, clearly provides important "islands" of habitat for several bird species that do not occur in the surrounding lowlands; to the extent that these forests may be threatened by climate change, it is an urgent priority to survey other mountains in the region to gain a more complete picture of the regional distributions of highland species.

LITERATURE CITED

Aleixo, A., F. Poletto, M. de Fátima Cunha Lima, M. Castro, E. Portes, and L. de Sousa Miranda. 2011. Notes on the vertebrates of northern Pará, Brazil: a forgotten part of the Guianan region, II. Avifauna. Boletim Museu Paraense Emílio Goeldi Ciências Naturais. 6: 11–65.

Chao, A. 1987. Estimating the population size for capture-recapture data with unequal catchability. Biometrics. 43: 783–791.

Chao, A., R.L. Chazdon, R.K. Colwell, and T-J. Shen. 2005. A new statistical approach for assessing similarity of species composition with incidence and abundance data. Ecology Letters. 8: 148–159.

Colwell, R.K. 2006. EstimateS: Statistical estimation of species richness and shared species from samples. Version 8. Persistent URL <purl.oclc.org/estimates>

Comité d'Homologation de Guyane. 2012. Bird List of French Guiana (version January 2012). Web site: http://gepog.pagesperso-orange.fr/CHG/

Mittermeier, J.C., K. Zyskowski, E.S. Stowe, and J.E. Lai. 2010. Avifauna of the Sipaliwini Savanna with insights into its biogeographic affinities. Bulletin of the Peabody Museum of Natural History. 51: 97–122.

O'Shea, B.J. 2005. Notes on birds of the Sipaliwini savanna and other localities in southern Suriname, with six new species for the country. Ornitología Neotropical. 16: 361–370.

O'Shea, B.J. 2008. Birds of the Konashen COCA, southern Guyana. In: Alonso, L.E., J. McCullough, P. Naskrecki, E. Alexander, and H.E. Wright (eds.). A Rapid Biological Assessment of the Konashen Community Owned Conservation Area, Southern Guyana. RAP Bulletin of Biological Assessment 51. Conservation International, Arlington VA. Pp. 63–68.

O'Shea, B.J., and S. Ramcharan. 2011. Avifauna of the Kwamalasamutu region, Suriname. In: O'Shea, B.J., L.E. Alonso, and T.H. Larsen (eds.). A Rapid Biological Assessment of the Kwamalasamutu region, Southwestern Suriname. RAP Bulletin of Biological Assessment 63. Conservation International, Arlington VA. Pp.131–143.

Ottema, O.H., J.H.J.M. Ribot, and A.L. Spaans. 2009. Annotated Checklist of the Birds of Suriname. WWF Guianas, Paramaribo.

Zyskowski, K., J.C. Mittermeier, O. Ottema, M. Rakovic, B.J. O'Shea, J.E. Lai, S.B. Hochgraf, J. de León, and K. Au. 2011. Avifauna of the easternmost tepui, Tafelberg in central Suriname. Bulletin of the Peabody Museum of Natural History. 52: 153–180.

Appendix 10.1. Cumulative Bird List, SE Suriname RAP, 8–29 March 2012. This list includes all species seen and heard during the SE Suriname RAP, including each of the main RAP camps (Juuru and Kasikasima), the Grensgebergte (Rock) satellite camp at 800 m, and the village of Palumeu. Taxonomy and nomenclature follow the most current version of the American Ornithologists' Union South American Checklist Committee's *Classification of the Bird Species of South America* (http://www.museum.lsu.edu/~Remsen/SACCBaseline.html).

Scientific name	English name	Juuru	Rock	Kasikasima	Palumeu	IUCN
Tinamidae						
Tinamus major	Great Tinamou	X	X	X	X	NT
Crypturellus cinereus	Cinereous Tinamou	X			X	
Crypturellus soui	Little Tinamou		X	X	X	
Crypturellus variegatus	Variegated Tinamou	X	X	X	X	
Crypturellus brevirostris	Rusty Tinamou			X		
Cracidae						
Penelope marail	Marail Guan	X		X		
Penelope jacquacu	Spix's Guan	X		X		
Pipile cumanensis	Blue-throated Piping-Guan	X				VU
Ortalis motmot	Variable Chachalaca	X	X		X	
Crax alector	Black Curassow	X		X		VU
Odontophoridae						
Odontophorus gujanensis	Marbled Wood-Quail	X	X	X	X	NT
Ardeidae						
Tigrisoma lineatum	Rufescent Tiger-Heron	X		X		
Zebrilus undulatus	Zigzag Heron	X			X	NT
Ardea cocoi	Cocoi Heron			X	X	
Pilherodius pileatus	Capped Heron			X	X	
Threskiornithidae						
Mesembrinibis cayennensis	Green Ibis	X		X	X	
Cathartidae						
Cathartes melambrotus	Greater Yellow-headed Vulture	X	X	X	X	
Sarcoramphus papa	King Vulture	X	X	X		
Pandionidae						
Pandion haliaetus	Osprey				X	
Accipitridae						
Elanoides forficatus	Swallow-tailed Kite	X	X		X	
Harpagus bidentatus	Double-toothed Kite	X		X	X	
Ictinia plumbea	Plumbeous Kite	X	X	X	X	
Buteogallus urubitinga	Great Black Hawk		X	X		
Geranoaetus albicaudatus	White-tailed Hawk			X		
Pseudastur albicollis	White Hawk	X				
Buteo brachyurus	Short-tailed Hawk		X			

Appendix 10.1. *continued*

Scientific name	English name	Juuru	Rock	Kasikasima	Palumeu	IUCN
Psophiidae						
Psophia crepitans	Gray-winged Trumpeter	X		X		
Rallidae						
Aramides cajaneus	Gray-necked Wood-Rail	X				
Anurolimnas viridis	Russet-crowned Crake		X		X	
Eurypygidae						
Eurypyga helias	Sunbittern	X				
Charadriidae						
Charadrius collaris	Collared Plover				X	
Scolopacidae						
Actitis macularius	Spotted Sandpiper			X	X	
Columbidae						
Columbina passerina	Common Ground Dove			X	X	
Columbina talpacoti	Ruddy Ground Dove				X	
Patagioenas speciosa	Scaled Pigeon				X	
Patagioenas plumbea	Plumbeous Pigeon	X	X	X	X	
Patagioenas subvinacea	Ruddy Pigeon	X	X	X	X	NT
Leptotila verreauxi	White-tipped Dove			X	X	
Leptotila rufaxilla	Gray-fronted Dove	X	X	X	X	
Geotrygon violacea	Violaceous Quail-Dove			X		
Geotrygon montana	Ruddy Quail-Dove	X		X		
Cuculidae						
Coccycua minuta	Little Cuckoo				X	
Piaya cayana	Squirrel Cuckoo	X	X	X	X	
Piaya melanogaster	Black-bellied Cuckoo			X		
Crotophaga ani	Smooth-billed Ani				X	
Strigidae						
Megascops watsonii	Tawny-bellied Screech-Owl	X		X		
Megascops guatemalae	Vermiculated Screech-Owl		X			
Lophostrix cristata	Crested Owl			X		
Pulsatrix perspicillata	Spectacled Owl	X				
Ciccaba virgata	Mottled Owl	X				
Glaucidium hardyi	Amazonian Pygmy-Owl	X				
Nyctibiidae						
Nyctibius grandis	Great Potoo				X	
Nyctibius leucopterus	White-winged Potoo		X	X		

table continued on next page

Appendix 10.1. *continued*

Scientific name	English name	Juuru	Rock	Kasikasima	Palumeu	IUCN
Caprimulgidae						
Lurocalis semitorquatus	Short-tailed Nighthawk			X		
Nyctidromus albicollis	Common Pauraque	X	X		X	
Caprimulgus nigrescens	Blackish Nightjar			X		
Hydropsalis climacocerca	Ladder-tailed Nightjar			X		
Apodidae						
Chaetura spinicaudus	Band-rumped Swift	X	X	X	X	
Chaetura chapmani	Chapman's Swift	X	X	X	X	
Aeronautes montivagus	White-tipped Swift			X		
Panyptila cayennensis	Lesser Swallow-tailed Swift	X	X			
Trochilidae						
Topaza pella	Crimson Topaz	X		X	X	
Phaethornis ruber	Reddish Hermit	X		X		
Phaethornis augusti	Sooty-capped Hermit		X			
Phaethornis bourcieri	Straight-billed Hermit	X	X	X		
Phaethornis superciliosus	Long-tailed Hermit	X		X		
Phaethornis malaris	Great-billed Hermit	X	X			
Heliothryx auritus	Black-eared Fairy	X		X		
Campylopterus largipennis	Gray-breasted Sabrewing	X	X	X	X	
Thalurania furcata	Fork-tailed Woodnymph	X	X	X		
Amazilia cf. *brevirostris*	Hummingbird sp.			X		
Amazilia viridigaster	Green-bellied Hummingbird		X			
Hylocharis sapphirina	Rufous-throated Sapphire	X				
Trogonidae						
Trogon melanurus	Black-tailed Trogon		X	X	X	
Trogon viridis	Green-backed Trogon	X		X	X	
Trogon violaceus	Guianan Trogon	X		X	X	
Trogon rufus	Black-throated Trogon			X		
Trogon collaris	Collared Trogon	X				
Alcedinidae						
Megaceryle torquata	Ringed Kingfisher	X		X	X	
Chloroceryle amazona	Amazon Kingfisher			X	X	
Chloroceryle americana	Green Kingfisher	X		X		
Chloroceryle inda	Green-and-rufous Kingfisher	X		X		
Chloroceryle aenea	American Pygmy Kingfisher	X		X		
Momotidae						
Momotus momota	Amazonian Motmot	X	X	X	X	

table continued on next page

Appendix 10.1. *continued*

Scientific name	English name	Juuru	Rock	Kasikasima	Palumeu	IUCN
Galbulidae						
Galbula albirostris	Yellow-billed Jacamar	X		X	X	
Galbula galbula	Green-tailed Jacamar				X	
Galbula dea	Paradise Jacamar	X	X	X	X	
Jacamerops aureus	Great Jacamar			X		
Bucconidae						
Notharchus macrorhynchos	Guianan Puffbird	X		X		
Bucco tamatia	Spotted Puffbird	X			X	
Bucco capensis	Collared Puffbird			X		
Malacoptila fusca	White-chested Puffbird	·		X		
Nonnula rubecula	Rusty-breasted Nunlet	X			X	
Monasa atra	Black Nunbird	X	X	X		
Chelidoptera tenebrosa	Swallow-winged Puffbird			X	X	
Capitonidae						
Capito niger	Black-spotted Barbet	X		X		
Ramphastidae						
Ramphastos tucanus	White-throated Toucan	X	X	X	X	
Ramphastos vitellinus	Channel-billed Toucan	X	X	X	X	
Selenidera culik	Guianan Toucanet	X		X		
Pteroglossus viridis	Green Aracari	X		X	X	
Pteroglossus aracari	Black-necked Aracari	X		X	X	
Picidae						
Picumnus exilis	Golden-spangled Piculet	X	X	X	X	
Veniliornis cassini	Golden-collared Woodpecker	X	X	X		
Piculus flavigula	Yellow-throated Woodpecker	X		X		
Colaptes rubiginosus	Golden-olive Woodpecker		X	X		
Celeus undatus	Waved Woodpecker	X		X		
Celeus elegans	Chestnut Woodpecker	X		X		
Celeus torquatus	Ringed Woodpecker	X		X		
Dryocopus lineatus	Lineated Woodpecker	X	X	X	X	
Campephilus rubricollis	Red-necked Woodpecker	X	X	X		
Campephilus melanoleucos	Crimson-crested Woodpecker	X		X	X	
Falconidae						
Micrastur ruficollis	Barred Forest-Falcon	X	X	X	X	
Micrastur gilvicollis	Lined Forest-Falcon			X	X	
Micrastur mirandollei	Slaty-backed Forest-Falcon	X				
Ibycter americanus	Red-throated Caracara	X	X	X	X	
Daptrius ater	Black Caracara			X	X	
Falco rufigularis	Bat Falcon			X	X	
Falco deiroleucus	Orange-breasted Falcon			X		NT

table continued on next page

Appendix 10.1. *continued*

Scientific name	English name	Juuru	Rock	Kasikasima	Palumeu	IUCN
Psittacidae						
Ara ararauna	Blue-and-yellow Macaw			X	X	
Ara macao	Scarlet Macaw	X		X		
Ara chloropterus	Red-and-green Macaw	X	X	X	X	
Ara severus	Chestnut-fronted Macaw	X				
Aratinga leucophthalma	White-eyed Parakeet			X	X	
Pyrrhura picta	Painted Parakeet	X		X	X	
Brotogeris chrysoptera	Golden-winged Parakeet	X	X	X	X	
Touit purpuratus	Sapphire-rumped Parrotlet			X		
Pionites melanocephalus	Black-headed Parrot	X		X		
Deroptyus accipitrinus	Red-fan Parrot	X		X	X	
Pyrilia caica	Caica Parrot	X		X		NT
Pionus menstruus	Blue-headed Parrot	X	X	X	X	
Pionus fuscus	Dusky Parrot	X		X	X	
Amazona amazonica	Orange-winged Parrot			X	X	
Amazona farinosa	Mealy Parrot	X	X	X	X	
Thamnophilidae						
Cymbilaimus lineatus	Fasciated Antshrike	X	X	X	X	
Frederickena viridis	Black-throated Antshrike	X		X		
Taraba major	Great Antshrike	X			X	
Thamnophilus murinus	Mouse-colored Antshrike	X		X	X	
Thamnophilus punctatus	Northern Slaty-Antshrike		X	X		
Thamnophilus amazonicus	Amazonian Antshrike	X		X		
Thamnomanes ardesiacus	Dusky-throated Antshrike			X		
Thamnomanes caesius	Cinereous Antshrike	X		X		
Isleria guttata	Rufous-bellied Antwren	X		X		
Epinecrophylla gutturalis	Brown-bellied Antwren	X		X		NT
Myrmotherula brachyura	Pygmy Antwren	X		X		
Myrmotherula surinamensis	Guianan Streaked-Antwren	X		X	X	VU
Myrmotherula axillaris	White-flanked Antwren		X	X	X	
Myrmotherula longipennis	Long-winged Antwren			X		
Myrmotherula menetriesii	Gray Antwren	X		X		
Herpsilochmus sticturus	Spot-tailed Antwren	X		X		
Herpsilochmus stictocephalus	Todd's Antwren	X	X	X		
Microrhopias quixensis	Dot-winged Antwren	X				
Hypocnemis cantator	Guianan Warbling-Antbird	X	X	X	X	NT
Terenura cf. spodioptila	Ash-winged Antwren	X		X		
Cercomacra cinerascens	Gray Antbird	X		X	X	
Cercomacra tyrannina	Dusky Antbird	X	X	X	X	
Myrmoborus leucophrys	White-browed Antbird	X			X	
Hypocnemoides melanopogon	Black-chinned Antbird	X			X	
Sclateria naevia	Silvered Antbird	X		X	X	
Percnostola rufifrons	Black-headed Antbird	X	X	X	X	

table continued on next page

Appendix 10.1. *continued*

Scientific name	English name	Juuru	Rock	Kasikasima	Palumeu	IUCN
Schistocichla leucostigma	Spot-winged Antbird	X				
Myrmeciza ferruginea	Ferruginous-backed Antbird	X		X		
Myrmornis torquata	Wing-banded Antbird	X		X		NT
Pithys albifrons	White-plumed Antbird	X		X		
Gymnopithys rufigula	Rufous-throated Antbird	X		X	X	
Hylophylax naevius	Spot-backed Antbird	X		X		
Willisornis poecilinotus	Common Scale-backed Antbird	X		X		
Conopophagidae						
Conopophaga aurita	Chestnut-belted Gnateater			X		
Grallariidae						
Grallaria varia	Variegated Antpitta	X	X	X		
Hylopezus macularius	Spotted Antpitta	X	X	X	X	
Myrmothera campanisona	Thrush-like Antpitta	X	X	X	X	
Formicariidae						
Formicarius colma	Rufous-capped Antthrush	X		X		
Formicarius analis	Black-faced Antthrush	X	X	X		
Furnariidae						
Sclerurus mexicanus	Tawny-throated Leaftosser	X				
Sclerurus rufigularis	Short-billed Leaftosser			X		
Sclerurus caudacutus	Black-tailed Leaftosser	X				
Deconychura longicauda	Long-tailed Woodcreeper			X		NT
Dendrocincla fuliginosa	Plain-brown Woodcreeper	X		X		
Glyphorhynchus spirurus	Wedge-billed Woodcreeper	X	X	X	X	
Dendrexetastes rufigula	Cinnamon-throated Woodcreeper	X			X	
Dendrocolaptes certhia	Amazonian Barred-Woodcreeper	X		X		
Dendrocolaptes picumnus	Black-banded Woodcreeper	X		X		
Hylexetastes perrotii	Red-billed Woodcreeper	X		X		
Xiphorhynchus pardalotus	Chestnut-rumped Woodcreeper	X	X	X		
Xiphorhynchus guttatus	Buff-throated Woodcreeper				X	
Campyloramphus procurvoides	Curve-billed Scythebill	X				
Lepidocolaptes albolineatus	Lineated Woodcreeper			X		
Xenops tenuirostris	Slender-billed Xenops	X				
Xenops minutus	Plain Xenops			X		
Philydor erythrocercum	Rufous-rumped Foliage-gleaner			X		
Philydor pyrrhodes	Cinnamon-rumped Foliage-gleaner	X			X	
Automolus ochrolaemus	Buff-throated Foliage-gleaner	X		X	X	
Automolus infuscatus	Olive-backed Foliage-gleaner	X		X		
Automolus rufipileatus	Chestnut-crowned Foliage-gleaner	X				
Synallaxis gujanensis	Plain-crowned Spinetail				X	
Synallaxis macconnelli	McConnell's Spinetail		X		X	

table continued on next page

Appendix 10.1. *continued*

Scientific name	English name	Juuru	Rock	Kasikasima	Palumeu	IUCN
Tyrannidae						
Tyrannulus elatus	Yellow-crowned Tyrannulet	X		X	X	
Myiopagis gaimardii	Forest Elaenia	X	X	X	X	
Myiopagis flavivertex	Yellow-crowned Elaenia	X				
Ornithion inerme	White-lored Tyrannulet			X		
Camptostoma obsoletum	Southern Beardless-Tyrannulet	X		X	X	
Corythopis torquatus	Ringed Antpipit	X		X		
Zimmerius acer	Guianan Tyrannulet	X	X	X		
Phylloscartes virescens	Olive-green Tyrannulet			X		
Mionectes oleagineus	Ochre-bellied Flycatcher			X		
Mionectes macconnelli	McConnell's Flycatcher	X	X	X		
Myiornis ecaudatus	Short-tailed Pygmy-Tyrant	X		X		
Lophotriccus vitiosus	Double-banded Pygmy-Tyrant	X		X		
Lophotriccus galeatus	Helmeted Pygmy-Tyrant	X		X		
Hemitriccus josephinae	Boat-billed Tody-Tyrant	X		X		
Hemitriccus zosterops	White-eyed Tody-Tyrant	X		X		
Todirostrum cinereum	Common Tody-Flycatcher		X	X	X	
Todirostrum pictum	Painted Tody-Flycatcher	X		X		
Tolmomyias assimilis	Yellow-margined Flycatcher	X	X	X		
Tolmomyias poliocephalus	Gray-crowned Flycatcher	X		X		
Platyrinchus saturatus	Cinnamon-crested Spadebill			X		
Platyrinchus platyrhynchos	White-crested Spadebill			X		
Myiophobus fasciatus	Bran-colored Flycatcher				X	
Terenotriccus erythrurus	Ruddy-tailed Flycatcher	X				
Hirundinea ferruginea	Cliff Flycatcher		X	X		
Contopus virens/sordidulus sp.	Eastern/Western Wood-Pewee		X			
Knipolegus poecilurus	Rufous-tailed Tyrant		X			
Legatus leucophaius	Piratic Flycatcher				X	
Myiozetetes cayanensis	Rusty-margined Flycatcher		X	X	X	
Myiozetetes luteiventris	Dusky-chested Flycatcher	X		X		
Pitangus sulphuratus	Great Kiskadee			X	X	
Conopias parvus	Yellow-throated Flycatcher		X	X	X	
Megarynchus pitangua	Boat-billed Flycatcher		X		X	
Tyrannus melancholicus	Tropical Kingbird		X	X	X	
Rhytipterna simplex	Grayish Mourner	X		X		
Sirystes sibilator	Sirystes			X		
Myiarchus tuberculifer	Dusky-capped Flycatcher	X		X		
Myiarchus ferox	Short-crested Flycatcher		X	X	X	
Ramphotrigon ruficauda	Rufous-tailed Flatbill			X		
Attila cinnamomeus	Cinnamon Attila				X	
Attila spadiceus	Bright-rumped Attila	X		X	X	

table continued on next page

Appendix 10.1. *continued*

Scientific name	English name	Juuru	Rock	Kasikasima	Palumeu	IUCN
Cotingidae						
Phoenicircus carnifex	Guianan Red-Cotinga			X		
Rupicola rupicola	Guianan Cock-of-the-rock	X	X	X		
Querula purpurata	Purple-throated Fruitcrow	X	X	X	X	
Perissocephalus tricolor	Capuchinbird	X		X		
Lipaugus vociferans	Screaming Piha	X	X	X	X	
Procnias albus	White Bellbird		X			
Xipholena punicea	Pompadour Cotinga	X				
Gymnoderus foetidus	Bare-necked Fruitcrow	X				
Pipridae						
Tyranneutes virescens	Tiny Tyrant-Manakin	X		X		
Corapipo gutturalis	White-throated Manakin	X	X	X		
Lepidothrix serena	White-fronted Manakin	X	X	X		
Manacus manacus	White-bearded Manakin	X		X	X	
Pipra pipra	White-crowned Manakin		X	X		
Pipra erythrocephala	Golden-headed Manakin	X	X	X		
Tityridae						
Schiffornis turdina	Brown-winged Schiffornis	X		X		
Pachyramphus marginatus	Black-capped Becard	X		X		
Pachyramphus minor	Pink-throated Becard	X				
Incertae sedis						
Piprites chloris	Wing-barred Piprites	X		X		
Vireonidae						
Cyclarhis gujanensis	Rufous-browed Peppershrike		X		X	
Vireolanius leucotis	Slaty-capped Shrike-Vireo	X	X	X		
Vireo olivaceus	Red-eyed Vireo			X		
Hylophilus thoracicus	Lemon-chested Greenlet	X		X	X	
Hylophilus sclateri	Tepui Greenlet		X			
Hylophilus muscicapinus	Buff-cheeked Greenlet	X		X		
Hylophilus ochraceiceps	Tawny-crowned Greenlet			X		
Hirundinidae						
Pygochelidon melanoleuca	Black-collared Swallow			X		
Atticora fasciata	White-banded Swallow			X	X	
Progne tapera	Brown-chested Martin				X	
Progne chalybea	Gray-breasted Martin				X	
Tachycineta albiventer	White-winged Swallow			X	X	
Hirundo rustica	Barn Swallow		X	X	X	

table continued on next page

Appendix 10.1. *continued*

Scientific name	English name	Juuru	Rock	Kasikasima	Palumeu	IUCN
Troglodytidae						
Microcerculus bambla	Wing-banded Wren	X		X		
Troglodytes aedon	House Wren				X	
Pheugopedius coraya	Coraya Wren	X	X	X	X	
Cantorchilus leucotis	Buff-breasted Wren	X		X	X	
Henicorhina leucosticta	White-breasted Wood-Wren		X			
Cyphorhinus arada	Musician Wren	X				
Polioptilidae						
Microbates collaris	Collared Gnatwren	X		X		
Ramphocaenus melanurus	Long-billed Gnatwren	X		X		
Polioptila plumbea	Tropical Gnatcatcher				X	
Turdidae						
Turdus fumigatus	Cocoa Thrush	X				
Turdus albicollis	White-necked Thrush	X	X	X		
Thraupidae						
Lamprospiza melanoleuca	Red-billed Pied Tanager	X	X	X		
Tachyphonus surinamus	Fulvous-crested Tanager	X		X		
Tachyphonus phoeniceus	Red-shouldered Tanager		X			
Lanio fulvus	Fulvous Shrike-Tanager	X		X		
Ramphocelus carbo	Silver-beaked Tanager	X	X	X	X	
Thraupis episcopus	Blue-gray Tanager		X	X	X	
Thraupis palmarum	Palm Tanager				X	
Tangara mexicana	Turquoise Tanager				X	
Tangara gyrola	Bay-headed Tanager		X			
Dacnis lineata	Black-faced Dacnis			X		
Dacnis cayana	Blue Dacnis			X		
Cyanerpes caeruleus	Purple Honeycreeper	X		X		
Cyanerpes cyaneus	Red-legged Honeycreeper		X	X		
Volatinia jacarina	Blue-black Grassquit			X	X	
Oryzoborus angolensis	Chestnut-bellied Seed-Finch	X	X	X		
Coereba flaveola	Bananaquit	X	X	X	X	
Incertae sedis						
Saltator grossus	Slate-colored Grosbeak	X	X	X		
Saltator maximus	Buff-throated Saltator	X		X	X	·
Emberizidae						
Zonotrichia capensis	Rufous-collared Sparrow		X			
Arremon taciturnus	Pectoral Sparrow	X		X	X	

table continued on next page

Appendix 10.1. *continued*

Scientific name	English name	Juuru	Rock	Kasikasima	Palumeu	IUCN
Cardinalidae						
Piranga flava	Hepatic Tanager		X			
Granatellus pelzelni	Rose-breasted Chat			X		
Caryothraustes canadensis	Yellow-green Grosbeak	X		X		
Periporphyrus erythromelas	Red-and-black Grosbeak			X		NT
Cyanocompsa cyanoides	Blue-black Grosbeak	X	X	X	X	
Parulidae						
Parula pitiayumi	Tropical Parula		X			
Phaeothlypis rivularis	Riverbank Warbler	X		X		
Icteridae						
Psarocolius viridis	Green Oropendola	X	X	X	X	
Psarocolius decumanus	Crested Oropendola				X	
Cacicus cela	Yellow-rumped Cacique			X	X	
Cacicus haemorrhous	Red-rumped Cacique			X		
Molothrus oryzivorous	Giant Cowbird				X	
Fringillidae						
Euphonia plumbea	Plumbeous Euphonia				X	
Euphonia violacea	Violaceous Euphonia			X	X	
Euphonia cyanocephala	Golden-rumped Euphonia		X			
Euphonia cayennensis	Golden-sided Euphonia	X	X	X		
313 spp.		196	103	233	133	

Chapter 11

Rapid Assessment Program (RAP) survey of small mammals in the Grensgebergte and Kasikasima region of Southeastern Suriname

Burton K. Lim and Hermando Banda

SUMMARY

A total of 39 species of small mammals (<1 kg) were documented during a biodiversity survey conducted from 9–24 March 2012 in a remote area of Southeastern Suriname. Taxonomic composition included 28 species of bats, 8 species of rats, and 3 species of opossums. The most common bat was the larger fruit-eating bat (*Artibeus planirostris*), which accounted for 38% of total captures. Although the capture rate for rats was substantially lower, as is typical of the Guiana Shield, spiny rats composed of 2 species (*Proechimys* spp.) were the commonest. For small opossums, there were 3 species documented by 1 individual each. Of the 3 sites sampled, the lowland sites were most similar with Upper Palumeu having the highest diversity of bats and Kasikasima having the highest abundance for bats. The highland site of Grensgebergte had the highest diversity and abundance for small non-volant mammals but the lowest for bats. This region of Southeastern Suriname has a mix of primary rainforest in a mosaic of lowland and highland habitats that supports diverse and different faunal communities of small mammals. The species composition was heterogeneous with no opossums shared among sites, whereas 25% of rats and just over 50% of bats were shared among sites. The most noteworthy records were the documentation of the poorly known water rat (*Nectomys rattus*) near the open granite outcrop of Grensgebergte.

INTRODUCTION

Small mammals (bats, rats, and opossums) that are less than 1 kg in body mass comprise approximately 75% of the mammalian species diversity in the Guianas (Lim et al., 2005). However, they are poorly known in comparison to the more charismatic and conspicuous larger species such as monkeys and cats. In Suriname, there are 194 species of mammals currently known from the country. Small mammals, in particular, are important for conservation because many are seed dispersers responsible for natural forest succession, pollinators of flowers, and controllers of insect populations through their foraging behavior and diet. High species diversity and relative abundance make small mammals an ideal group for rapid assessment program (RAP) surveys and long term monitoring. This is particularly important for regions such as the Grensgebergte area that have not been surveyed for biodiversity and conservation purposes (Husson, 1978).

METHODS

(1) The first study area (Site 1) was a camp on the Upper Palumeu River along a trail to Brazil used by the local Amerindians (N 2.47705°, W 55.62941°, 234 m elevation). It was situated in rainforest on rolling terrain and surveyed for 6 nights from 9–14 March 2012. (2) The second area (Site 2) was a granite outcrop in the Grensgebergte mountain range (N 2.52667°, W 55.77018°, 778 m) in montane forest. It was surveyed for 3 nights from 15-17 March 2012. (3) The third area (Site 4; the third site was surveyed primarily for aquatic organisms) was near Kasikasima across from an Amerindian village on the Palumeu River (N 2.97741°, W 55.38479°, 210 m). It was situated in rainforest on rolling terrain and surveyed for 7 nights from 20-26 March 2012. Mist nets were set at the base of Kasikasima mountain (N2.97741°, W55.40770°, 277 m) approximately 3 km west of Site 4 on the last evening of sampling.

Survey methods for small mammals followed standard protocols outlined in the report of an earlier RAP survey of the Kwamalasamutu region of Suriname (Lim and Joemratie, 2011). Sherman live traps of 2 sizes (23 × 8 × 9 cm and 35 × 12 × 14 cm) were used for sampling the terrestrial and arboreal rats, and small opossums. A maximum of 120 traps were set approximately 5 meters apart both on the ground and in trees along a transect within the forest. For bats, mist nets were usually set in pairs (12 × 2.6 and 6 × 2.6 m) approximately 100 meters apart within the forest understory across transect trails, over creeks, in swamps, near tree fall gaps, and by rocky outcrops. A maximum of 21 mist nets were set and opened from approximately 18:00 to 24:00 h.

Small mammals caught were preliminarily identified in the field and up to 2 individuals per species per night were kept as a representative collection of species diversity. All other individuals were released. Voucher specimens were prepared as dried skins with carcasses temporarily preserved in ethanol for later cleaning of the skulls and skeletons in a dermestid beetle colony at the Royal Ontario Museum (ROM), or as whole animals fixed in 10% formalin with later long-term storage in 70% ethanol. This will enable examination of both osteology and soft anatomy for morphological study and confirmation of field identifications. Tissue samples of liver, heart, kidney, and spleen were preserved in 95% ethanol for later storage in a −80°C ultra-cold freezer for molecular study of genetic variation and verification using DNA barcoding (Borisenko et al. 2008).

A reference collection of voucher specimens will be deposited at the University of Suriname's National Zoological Collection of Suriname, and the Royal Ontario Museum as documentation of the biodiversity of mammals in South-eastern Suriname and available for study by the scientific community. They will also contribute to ongoing research projects on the evolution of mammals in the Neotropics and comparison of community ecology in the Guianas.

Species diversity and relative abundance data was analyzed using EstimateS version 8.2 (Colwell 2009). Biodiversity measures included the calculation of 7 species richness estimators, 4 diversity indices, 8 shared species similarity indices, and species accumulation curves with asymptote functions.

RESULTS

Preliminary field identifications documented 39 species of small mammals represented by 366 individual captures, of which 189 were kept as voucher specimens and 177 individuals were released. More specifically, 28 species of bats were represented by 345 individuals, 8 species of rats and mice were represented by 18 individuals, and 3 species of opossums were represented by 3 individuals (Appendix 11.1). One bat (*Molossus rufus*) and one opossum (*Gracilianus emiliae*; Figure 11.1) were caught by hand and included in our report but not used in the analysis of biodiversity measures. The richness estimators ranged from 31 to 41 species of bats and 14 to 22 species of rats and opossums expected with the methods used (Table 11.1). In addition, a gray four-eyed opossum (*Philander opossum*) was seen climbing up a tree in a swampy area on the trail to Kasikasima Mountain but not collected.

The commonest species of bat (larger fruit-eating bat, *Artibeus planirostris*; Figure 11.2) represented over ⅓ of the total captures in mist nets. It was more than twice as frequently caught as the next most abundant species (moustached bat, *Pteronotus parnellii*; Figure 11.3) and was documented at all 3 sites as were 2 other species (Seba's short-tailed fruit bat, *Carollia perspicillata* [Figure 11.4], and round-eared

bat, *Lophostoma silvicolum* [Figure 11.5]). By contrast, the 3 species of opossums were caught only once, each at one of the 3 different sites. For rats and mice, the Guianan terrestrial spiny rat (*Proechimys guyannensis*) was the most common with 5 captures but it was caught at only Site 2 in

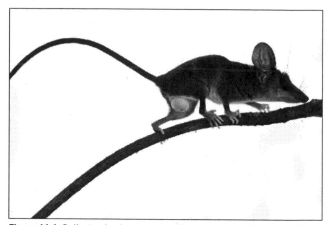

Figure 11.1. Delicate slender opossum (*Marmosops parvidens*) caught at the Upper Palameu site by Piotr Naskrecki. Photo by Piotr Naskrecki.

Figure 11.2. Larger fruit-eating bat (*Artibeus planirostris*). Photo by Burton Lim.

Figure 11.3. Common moustached bat (*Pteronotus parnellii*). Photo by Burton Lim.

Table 11.1. Summary of the small mammal RAP survey data in Southeastern Suriname. The data is separated into the 3 sampling sites: Upper Palumeu (UP), Grensgebergte (GR), and Kasikasima (KA). The observed data are summarized from Appendix 11.1. Richness estimators, asymptote function, and diversity indices were computed using EstimateS (Colwell 2009).

Variable	Bats					Small non-volant mammals			
	UP	GR	KA	Total		UP	GR	KA	Total
Observed data:									
Individuals	112	14	218	334		2	14	4	20
Species	21	5	19	27		2	7	4	10
Trap-nights	86	30	117	223		518	272	837	1727
Richness estimators:									
ACE	24	8	24	34		3	13	10	22
ICE	28	11	26	34		3	12	9	20
Chao 1	25	6	23	41		3	10	10	21
Chao 2	23	6	22	31		3	8	9	16
Jack 1	27	7	25	34		4	10	7	18
Jack 2	28	8	27	37		5	11	10	21
Bootstrap	24	6	22	31		8	8	5	14
Average richness	26	7	24	35		4	10	9	19
Asymptote function:									
MMMeans	29	12	27	33		2	16	6	32
Diversity indices:									
Alpha	8	3	5	7		12974	6	29192	10
Shannon	8	1	2	2		1	2	1	2
Shannon Exponential	11	4	6	9		2	6	4	9
Simpson	8	4	4	5		1	7	1	11

Table 11.2. Shared species and similarity indices between sampling sites (Upper Palumeu – UP; Grensgebergte – GR; and Kasikasima – KA) for small mammals surveyed in Southeastern Suriname as computed using EstimateS (Colwell, 2009).

Group	First Sample	Second Sample	Shared Species	Jaccard Classic	Sorensen Classic	Chao-Jaccard-Raw Abundance-based	Chao-Jaccard-Est Abundance-based	Chao-Sorensen-Raw Abundance-based	Chao-Sorensen-Est Abundance-based	Morisita-Horn	Bray-Curtis
Bats	UP	GR	4	0.182	0.308	0.253	0.282	0.404	0.44	0.37	0.127
	UP	KA	14	0.538	0.7	0.868	0.883	0.929	0.938	0.835	0.509
	GR	KA	3	0.143	0.25	0.353	0.357	0.522	0.526	0.504	0.06
Small non-volant mammals	UP	GR	0	0	0	0	0	0	0	0	0
	UP	KA	1	0.2	0.333	0.2	0.274	0.333	0.43	0.333	0.333
	GR	KA	1	0.1	0.182	0.1	0.124	0.182	0.22	0.154	0.111

the Grensgebergte Mountains. The closely related Cuvier's terrestrial spiny rat (*P. cuvieri*) was one of the next commonest species with 3 individuals caught. These large rats (ca. 500 g) are prey for many top-level predators such as cats

Figure 11.4. Seba's short-tailed fruit bat (*Carollia perspicillata*). Photo by Burton Lim.

Figure 11.5. Round-eared bat (*Lophostoma silvicolum*). Photo by Burton Lim.

Figure 11.6. Water rat (*Nectomys rattus*) caught at the Grensgebergte site. Photo by Trond Larsen.

and snakes. The only rat that was caught at all 3 sites was the arboreal rice rat *Oecomys auyantepui*. The water rat *Nectomys rattus* (Figure 11.6) was also captured 3 times, but all were from the Grensgebergte Mountains site.

In terms of study sites, the first site on the Upper Palumeu documented 25 species of small mammals that were represented by 116 individuals including 22 species of bats (113 individuals), 2 species of rats (2 individuals), and 1 species of opossum (1 individual). The second site on Grensgebergte documented 12 species of small mammals represented by 28 individuals including 5 species of bats (14 individuals), 6 species of rats (13 individuals), and 1 species of opossum (1 individual). The third camp at Kasikasima documented 23 species of small mammals represented by 222 individuals including 19 species of bats (218 individuals), 3 species of rats (3 individuals), and 1 species of opossum (1 individual).

The diversity indices indicated that Upper Palumeu was the most diverse site for bats and Grensgebergte was the least diverse for bats (Table 11.1). In contrast, Grensgebergte was the most diverse site for small non-volant mammals. Upper Palumeu and Kasikasima had the most shared species of bats and the highest similarity indices (Table 11.2). The numbers were low but Upper Palumeu and Kasikasima also had slightly higher similarity indices than the other 2 site comparisons for small non-volant mammals.

The species accumulation curves for bats are quite distinct from each other with Grensgebergte being the most different with no overlap in number of species or individuals when compared to the other 2 sites (Figure 11.7). The asymptote function of the curve also levels off at half the values for Upper Palumeu and Kasikasima (Table 11.1). For small non-volant mammals, the accumulation curve for Grensgebergte is again distinct from the other 2 sites but with more numbers of species and individuals (Figure 11.8), which is reflected in its higher asymptote function (Table 11.1). The asymptote function calculated for the total survey of bats indicated 81% completeness of survey (observed/estimated; 27/33) based on the methods used. The completeness of survey for small non-volant mammals was much lower at 34%.

DISCUSSION

Each of the 3 sites surveyed for small mammals during the Southeastern Suriname RAP had a different faunal composition in terms of species diversity and relative abundance. The highland site of Grensgebregte was the most distinctive, whereas the lowland sites of Upper Palumeu and Kasikasima were the most similar. Although not numerous in terms of diversity or abundance, half of the 8 species of rats documented during the survey were found only at Grensgebergte. This indicates that the montane habitats support a unique community ecology of non-volant small mammals that is different from the surrounding lowland regions. Although the number of species was similar to an earlier RAP to the

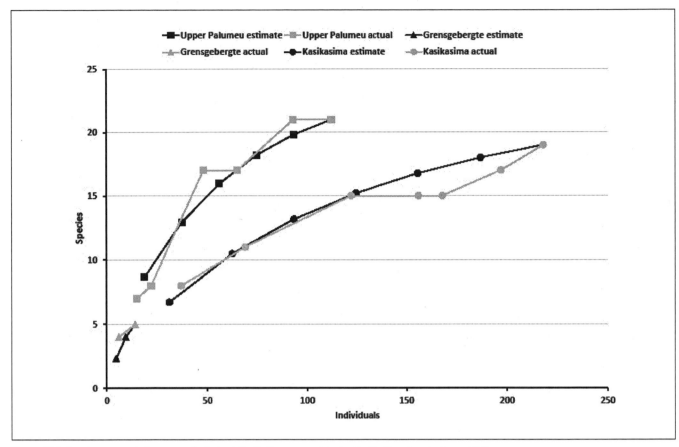

Figure 11.7. Species accumulation curves for bats at the 3 sites surveyed during the Southeastern Suriname RAP.

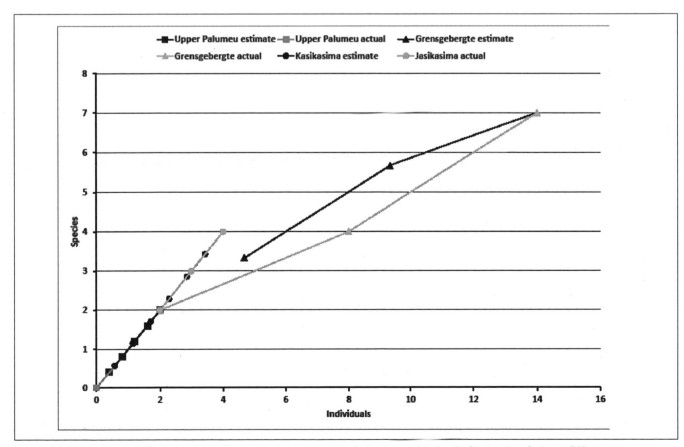

Figure 11.8. Species accumulation curves for small non-volant mammals at the 3 sites surveyed during the Southeastern Suriname RAP.

Kwamalasamutu region of Suriname, the number of individuals caught in the Palumeu region was almost 10-fold less (18:149). However, the Kwamalasamutu RAP was anomalous in that the rat abundance was unusually high for the Guiana Shield (Lim and Joemratie 2011).

The species diversity of bats was highest at Upper Palumeu, which suggests that this forested area is in pristine condition and is representative of primary lowland rainforest in the Guiana Shield of South America. In contrast, Kasikasima had nearly as high species diversity but had exceptionally high abundance (over twice as many as the other sites) for 2 species of bats. The first species (Parnell's moustached bat; *Pteronotus parnellii*) was very common in the rocky outcrops at the base of Kasikasima Mountain, approximately 3 km inland from our basecamp situated on the right bank of the Palumeu River. The crevices and overhangs of the granite boulders were used as day roosts because our nets caught primarily this species in the early evening. The second species (larger fruit-eating bat; *Artibeus planirostris*) was very abundant in the forest surrounding our campsite, which was across the river from an Amerindian settlement. The supply of fruits necessary to maintain high populations of these large fruit-eating bats may be the ephemeral result of coincidental masting of fruiting trees in the area or perhaps a more commensal association with human cultivation activities. A similarly high abundance of *Artibeus planirostris* was found at Werehpai on a previous RAP survey to the Kwamalasamutu region of Suirname (Lim and Joemratie 2011). However, this result was attributed in part to the established trail system to the petroglyph caves that were acting as flyways and facilitated the netting of bats. Although the area was previously used for farming, there was not the persistant human presence as found at Kasikasima. Nonetheless, in some habitats *Artibeus planirostris* can be the dominant species of bat with a major influence on seed dispersal and forest succession.

In comparison to other small mammal studies in Suriname, the results of our survey in the Grensgebergte area was similar to Kwamalasamutu (Lim and Joemratie 2011) with *Artibeus planirostris* being the most common species of bat. Among sites between these RAP surveys, species diversity and relative abundance were similar for the lowland sites of Upper Palumeu and Werehpai. By contrast, the short-tailed fruit bat *Carollia perspicillata* was the commonest at Bakhuis (Lim 2009) and Brownsberg (Lim et al. 2005) in more disturbed areas in the north.

In terms of small non-volant mammals, Grensgebergte was similar to Bakhuis with species of terrestrial spiny rats (*Proechimys* spp.) being most common, whereas Kwamalasamutu and Brownsberg had other species as the commonest. For species diversity and relative abundance, Kwamalasamutu was unusually high for the Guiana Shield (Lim and Joemrate 2011) with this Grensgebergte survey more typical of the region.

One of the more interesting discoveries was the capture of 3 individuals of the water rat *Nectomys rattus* (Figure 11.6) from the Grensgebergte Mountains site. This is a water-adapted species with webbing on its hindfeet. All were caught in habitat with ephemeral water sources near the tops of the mountain range.

CONSERVATION RECOMMENDATIONS

Grensgebergte was the most remote and unexplored of the three sites surveyed for small mammals. Although it was sampled only for three days, there was relatively high species diversity for rats in comparison to lowland sites. The surrounding mountain ranges in southeast Suriname warrant more biological study to ascertain and uncover the unique faunal and floral community that resides in this pristine tropical environment. These low-elevation mountains can function as important corridors of biodiversity between protected areas in neighboring countries of the Guiana Shield region.

The Upper Palumeu had the highest species diversity for bats indicating its relatively undisturbed habitat. Although there is a camp and trail connecting Suriname and Brazil, it is only used infrequently by the Amerindians for travel between villages in the area. This rolling terrain acts as an ecological buffer for the mountain ranges that are interspersed within the lowland rainforest.

Kasikasima had the highest relative abundance of bats and in particular a large proportion of the fruit-eating bat species *Artibeus planirostris*. This indicates a healthy supply of fruiting trees that can sustain a large population of this frugivorous animal. There is a nearby village and ecotourist facilities but the impact on the surrounding forest seems minimal and the trail to Kasikasima Mountain is not regularly maintained. Incorporating all three sites or areas into a multi-use nature reserve will benefit the connectivity of surrounding protected areas, and the conservation and promotion of biodiversity.

LITERATURE CITED

Borisenko, A.V., B.K. Lim, N.V. Ivanova, R.H. Hanner, and P.D.N. Hebert. 2008. DNA barcoding in surveys of small mammal communities: a field study in Suriname. Molecular Ecology Resources, 8:471–479.

Colwell, R.K. 2009. EstimateS: statistical estimation of species richness and shared species from samples. Version 8.2. User's guide and application published at: http://purl.oclc.org/estimates.

Husson, A.M. 1978. The mammals of Suriname. Zoological Monographs. Rijksmuseum Natural History. 2: 1–569.

Lim, B.K. 2009. Environmental assessment at the Bakhuis bauxite concession: small-sized mammal diversity and abundance in the lowland humid forests of Suriname. The Open Biology Journal. 2:42–53.

Lim, B.K., M.D. Engstrom, H.H. Genoways, F.M. Catzeflis, K.A. Fitzgerald, S.L. Peters, M. Djosetro, S. Brandon, and S. Mitro. 2005. Results of the Alcoa Foundation-Suriname expeditions. XIV. Mammals of Brownsberg Nature Park, Suriname. Annals of Carnegie Museum. 74: 225–274.

Lim, B.K., M.D. Engstrom, and J. Ochoa G. 2005. Mammals. *In:* T. Hollowell and R. P. Reynolds (eds.). Checklist of the terrestrial vertebrates of the Guiana Shield. (Bulletin of the Biological Society of Washington, 13: 77–92.

Lim, B.K., and S. Joemratie. 2011. Rapid Assessment Program (RAP) survey of small mammals in the Kwamalasamutu region of Suriname. *In:* B.J. O'Shea, L.E. Alonso, and T.H. Larsen (eds.). A rapid biological assessment of the Kwamalasamutu Region, Suriname (Conservation Internationa., RAP Bulletin of Biological Assessment 63. Pp. 144–149.

Appendix 11.1. Preliminary checklist of 39 species of small mammals documented from 3 sites during the RAP of the Grensgebergte area and number of individuals captured.

Asterisk (*) indicates 2 species that were caught by hand and not by traps or nets.

Species	Upper Palumeu	Grensgebergte	Kasikasima	Total
Opossums (3 species)				
*Marmosops parvidens**	1			1
Marmosa murina		1		1
Monodelphis brevicaudatus			1	1
Subtotal	1	1	1	3
Rats (8 species)				
Neacomys dubosti	1			1
Oecomys auyentepui	1	1	1	3
Oecomys rutilus		1		1
Nectomys rattus		3		3
Hylaeamys megacephalus			1	1
Proechimys cuvieri		2	1	3
Proechimys guyannensis		5		5
Rhipidomys nitela		1		1
Subtotal	2	13	3	18
Bats (28 species)				
Ametrida centurio	3			3
Anoura caudifer			1	1
Artibeus bogotensis		6		6
Artibeus lituratus	1		2	3
Artibeus obscurus	18		12	30
Artibeus planirostris	29	4	99	132
Carollia brevicauda	2		1	3
Carollia perspicillata	2	2	16	20
Chiroderma villosum	1			1
Chrotopterus auritus	3		2	5
Desmodus rotundus	1			1
Glossophaga soricina			1	1
Lionycteris spurrelli	3		1	4
Lonchophylla thomasi	4		5	9
Lophostoma silvicolum	3	1	4	8
Mesophylla macconnelli	1			1
Micronycteris megalotis			1	1
Mimon crenulatum			1	1
*Molossus rufus**	1			1

table continued on next page

Appendix 11.1. continued

Species	Upper Palumeu	Grensgebergte	Kasikasima	Total
Phyllostomus elongatus	4		5	9
Phyllostomus hastatus	3			3
Platyrrhinus helleri	1	1		2
Pteronotus parnellii	17		47	64
Rhinophylla pumilio	4		8	12
Sturnira lilium			5	5
Sturnira tildae	2		5	7
Trachops cirrhosus	9		2	11
Uroderma bilobatum	1			1
Subtotal	113	14	218	345
Total	116	28	222	366

Chapter 12

A survey of the large ground dwelling mammals of the Upper-Palumeu river region

Krisna Gajapersad

SUMMARY

During the survey of large and medium-sized mammals of the Kasikasima and Upper Palumeu regions we recorded 18 species. Camera traps were the most important tools for the survey, but direct observations were made and tracks, scat and scratch marks were also recorded. Important species such as Jaguar, Tapir and Giant river otter were recorded. All these species fulfil important roles in the ecosystem such as controlling populations and seed dispersers. The occurrence of a high diversity of large and medium-sized mammals in the surveyed area indicates that the ecosystem is healthy and relatively pristine. Southeastern Suriname is very important for large mammal species, because the area encompasses vast tracts of pristine forest and rivers. In fact, there are few places left on earth which are as pristine as Southeastern Suriname. Hunting and habitat destruction were identified as the greatest potential threats for this area.

INTRODUCTION

Large and medium-sized mammals are the most important targets for commercial and subsistence hunting for meat, skins and other parts of the body. Furthermore, large and medium-sized mammals play important roles in the ecosystem as agents of seed dispersal, animal population regulation, and habitat maintenance and modification. These important functions of large and medium-sized mammals in the ecosystem make them important indicators of the functioning of ecosystems and human pressure (Lahm and Tezi 2003). Historically humans have used animals for food and a variety of other uses (Leader-Williams et al. 1990, Milner-Gulland et al. 2001). Examples all over the world show the effects of overhunting from humans causing population decreases and extinction (Diamond 1989). Overexploitation was almost certainly responsible for historical extinctions of some large mammals and birds (Turvey and Risley 2006). Large mammals have slow reproductive rates, long development and growth times and large requirements of food and habitat, which make them more sensitive to hunting and

habitat destruction (Purvis et al. 2000, Cardillo et al. 2005). Today, about 2 million people depend on wild meat for food or trade (Fa et al. 2002, Milner-Gulland et al. 2003), yet the majority of hunting is unsustainable (Robinson and Bennett 2004, Silvius et al. 2005).

Mammals as a group provide the main protein source for native Amazonian communities. Indigenous peoples have lived in Amazonia for tens of thousands of years (Redford 1992) and many, including the indigenous groups in Suriname, still remain within the forest and hunt mammals actively. In areas where they have been hunted, abundance of large mammals has decreased (Peres 1990; Cullen et al. 2000; Hill et al. 2003). Unmanaged hunting is commonplace in the Amazon and is depleting game populations, often to levels so low that local extinctions are frequent and the critical ecological roles performed by mammals are lost (Redford 1992; Bodmer et al. 1994). Overhunting then becomes a double-edged threat: to the biodiversity of the tropics and to the people that depend on those harvests for food and income. Little information is available on the occurrence, spatial variability in species richness, and sensitivity to hunting and other disturbances of medium and large mammals in Suriname, especially in the remote southern part of the country.

METHODS AND STUDY SITES

We surveyed medium- and large-bodied mammals by means of three main methods: camera trapping, searching for scat and animal tracks, and making visual and aural observations.

Camera traps were set approximately 500 meters apart along trails which were cut during the camera trap set up, on game trails and along an existing tourist trail. The Reconyx RM45 Rapidfire, Reconyx PC800 Hyperfire, Reconyx HC500 Hyperfire and the Snapshot Sniper are the four different camera models which were used for the survey. The camera traps operated day and night, photographing all ground-dwelling mammals and birds that walked in front of them. Camera traps were attached to trees approximately 30 cm above the forest floor. At the Upper Palumeu site

(RAP Site 1) 12 camera traps were set up along one trail of approximately 5 kilometers that was cut during the set up of the camera traps. Camera traps were set up at locations where tracks and scat were found and along game trails. At Kasikasima (RAP Site 4) 14 camera traps were set up along two trails. Eight camera traps were setup along the tourist trail that leads from the tourist camp along the Palumeu river to the Kasikasima mountain and six camera traps were set up along an old trail, which leads from the second RAP camp to the Kasikasima mountain. The second trail had to be re-opened because it had not been used by local people for a long time. Cameras were placed in different habitats at each of the study sites. At the Upper Palumeu site 8 camera traps were set up in terra firme, two in swamp forest and two along dry creek beds. At the Kasikasima site 11 camera traps were set up in terra firme, two in swamp forest along creeks and one in a liana forest along the mountain. Elevations of camera trapping points at the Upper Palumeu site ranged between 278 and 407 meters and at the Kasikasima site the elevation ranged between 211 and 477 meters.

Camera trapping pictures were identified to species and independent pictures were used as single occurrences for analysis of the data obtained with the camera traps. Any photographs taken at least 30 minutes apart and photographs of different species and individuals were used as independent samples for analysis (O'Brien et al. 2003). Species accumulation curves and biodiversity indices were calculated with the software program EstimateS 8.2.0 (Colwell 2009).

Occurrences from photographs were compared with the nonparametric richness estimator Chao 1 among the Upper Palumeu and Kasikasima areas (Magurran 2004). For the analysis one day was split up in 2 camera trapping occasions.

Visual and aural observations, tracks, scat and scratch marks of large cats were recorded when walking the trails to set up and pick up the camera traps. Primates are not included in this chapter, since they were surveyed by a separate primatologist during the same expedition.

RESULTS AND DISCUSSION

During the RAP we recorded 18 species of medium- and large-bodied ground dwelling mammals (Table 12.1). We recorded 11 mammal species at the Upper Palumeu site (RAP Site 1) in a total of 99 camera trapping days with 11 camera traps. At the Kasikasima site (RAP Site 4) we found 15 mammal species in a total of 209 camera trapping days with 13 camera traps.

The large caviomorph rodents were the most frequently photographed by the camera traps; this group was assumed to be the most common group of medium- and large-bodied non-volant mammals in the area. The rodent species most frequently photographed by the camera traps were Paca (*Cuniculis paca*), Red-rumped Agouti (*Dasyprocta leporina*) (see page 32) and Red-acouchy (*Myoprocta acouchy*).

Table 12.1. Species of large and medium sized mammals of the Upper Palumeu (RAP Site 1) and Kasikasima (RAP Site 4).

Scientific name	Common name	Detection method	Site	IUCN Category
Cuniculus paca	Spotted Paca	CT	UP, KK	LC
Dasyprocta leporina	Red-rumped agouti	CT, O	UP, KK	LC
Dasypus kappleri	Great long-nosed armadillo	CT	KK	LC
Dasypus novemcinctus	Nine-banded armadillo	CT	UP, KK	LC
Didelphis marsupialis	Common opossum	CT	KK	LC
Eira barbara	Tayra	CT, O	UP, KK	LC
Leopardus pardalis	Ocelot	CT	UP, KK	LC
Leopardus wiedii	Margay	CT, O	KK	NT
Mazama americana	Red brocket deer	CT, O	UP, KK	DD
Mazama gouazoubira	Grey brocket deer	CT	UP	LC
Myoprocta acouchy	Red acouchi	CT	UP, KK	LC
Panthera onca	Jaguar	T	UP, KK	NT
Pecari tajacu	Collared peccary	CT	UP, KK	LC
Pteronura brasiliensis	Giant river otter	O	Makrutu	EN
Puma concolor	Puma	CT	KK	LC
Sciurus aestuans	Guianan squirrel	CT	KK	LC
Tayassu pecari	White-lipped peccary	O	KK	NT
Tapirus terrestris	Brazilian tapir	T	UP	VU

CT = Camera trap; O = Direct Observation; T = Tracks; UP = Upper Palumeu; KK = Kasikasima
IUCN Red List Categories: LC = Least concern; NT = Near threatened; DD = Data deficient; EN = Endangered; VU = Vulnerable

Of the six species of cats known to occur on the Guiana Shield, the Jaguar (*Panthera onca*), Puma (*Puma concolor*) (see page 32), Ocelot (*Leopardus pardalis*) (see page 32) and Margay (*Leopardus wiedii*) (see page 31) were found during the survey. Ocelot was the most frequent recorded cat species during this survey and is common in the area. Tracks and scratch marks of Jaguar were found at the Upper Palumeu site. Tracks of Jaguar were also found in the area of the foot of the Kasikasima rock. During the survey the camera traps did not take pictures of Jaguar. The Jaguar, at top of its food chain, is an important keystone species, because it has a regulating function in the ecosystem where it occurs. The Jaguar preys on herbivorous and granivorous mammal populations and thus it has a positive effect on the plant communities in the ecosystem. Puma was photographed only once by the camera traps, on the frequently visited tourist trail that leads to the Kasikasima rock. It is very likely that the Puma also occurs in the Upper Palumeu area, but was only recorded in the Kasikasima area because the trail at Kasikasima is used frequently by these large cats. In both areas that were surveyed we recorded Red-brocket deer (*Mazama americana*) (see page 32). The Collared Peccary (*Pecari tajacu*) was recorded at both areas and was especially abundant in the Upper Palumeu area. The White-lipped Peccary (*Tayassu pecari*) was not recorded by the camera traps but a group was observed by the local workers during the survey at the Kasikasima area. The Giant river otter (*Pteronura brasiliensis*) was recorded at the mouth of the Makrutu creek. This species is an important indicator species, because they are very sensitive to changes in the aquatic ecosystem. The presence of Giant river otter in this area indicates that the aquatic ecosystem is still healthy. This species is also important for ecotourism, because it is very charismatic and easy to spot on the river. Two armadillo species were found during the RAP, including Great Long-nosed Armadillo (*Dasypus kappleri*) and the Nine-banded Armadillo (*Dasypus novemcinctus*). The Giant armadillo (*Priodontes maximus*) was not recorded during the RAP, but it is likely that it is present in both areas, because of the large burrows and sightings by local people. Surprisingly, the Brazilian tapir (*Tapirus terrestris*), which we expected to be quite common, was only recorded by tracks at one location at the Upper Palumeu site. A number of ground-dwelling bird species important as a food source for the local people were recorded by the camera traps and observed during the RAP, including Black curassow (*Crax alector*), Grey-winged trumpeter (*Psophia crepitans*) (see page 30) and Great tinamou (*Tinamus major*).

The species accumulation curves show that many more species likely occur at both sites than were found in this study, which is not surprising given the short sampling period (Figure 12.1). This is confirmed by the Chao 1 diversity estimator, which predicts that additional species would be detected by the camera traps with greater sampling effort at both sites (Figure 12.2).

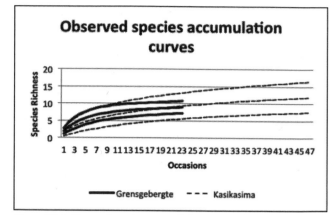

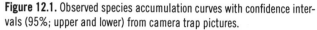

Figure 12.1. Observed species accumulation curves with confidence intervals (95%; upper and lower) from camera trap pictures.

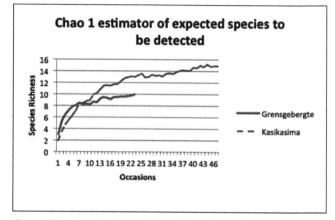

Figure 12.2. Chao 1 estimator of expected species to be detected by camera traps at each RAP survey site.

CONSERVATION RECOMMENDATIONS

It is likely that the medium- and large-bodied mammals found during this RAP survey are distributed throughout southern Suriname, since the medium and large mammal fauna (with the exception of primates) is generally consistently distributed throughout the Amazon Basin. The species recorded during this RAP were expected, although some species appear to occur at lower abundances than we expected. The camera traps were a very important tool during this RAP survey, because most of the large mammals were recorded by the camera traps and not by other observation methods. More species were recorded in the Kasikasima area than in the Upper Palumeu, but this may only reflect differences in sampling effort, with more camera trapping days around Kasikasima. Furthermore, the loud sound of the helicopter that flew daily in the Upper Palumeu area during the survey may have reduced mammal observations at this site.

The Upper Palumeu area is quite pristine, and is visited only occasionally by the local people while they travel from the villages in southern Suriname to villages in Brazil. Hunting does not seem to pose a threat for the large and medium-sized mammals in the Upper Palumeu area, because the area is so remote and very inaccessible. We found no signs of hunting or any other human disturbance. In the Kasikasima area there are more human activities than in the Upper Palumeu area, such as ecotourism and some hunting by a small number (10 - 20) of people that live in Kampu. Kampu is a small settlement along the Palumeu river a few kilometers downstream of the Papadron rapids. Currently the greatest potential threat for large and medium-sized mammals at Kasikasima would be hunting, but the results from the RAP survey show that hunting pressure is currently low.

The results of this RAP survey cannot provide an accurate indication of the population status of the different large mammal species, because we were not able to calculate species densities or relative abundance from the data that was gathered during the short timeframe of the survey. Nevertheless, the presence of species sensitive to hunting and disturbance such as Jaguar, Puma, Tapir, curassows and large primates also suggests that hunting pressure is low, and that the important ecological processes maintained by these species, such as seed dispersal and population regulation, remain intact.

Hunting is probably limited by reduced river access to some areas in the dry season, and more generally by distance from Palumeu and other villages. The absence of a market and the concentration of the indigenous people in Palumeu both reduce hunting pressure on large vertebrates in the region as a whole. The extensive surrounding forest acts as a source to offset local population depletion due to hunting. Nevertheless, the most significant current threat to medium- and large-bodied mammals in the area is hunting from Palumeu village. This could change if plans to build a road from northern to southern Suriname move ahead. A road would make the area accessible and hunting and habitat destruction would become important threats for large terrestrial vertebrates. Further plans to increase hydro energy capacity of the existing hydro lake in Brokopondo by diverting water from the Tapanahony river to the Jai creek could be a threat if the project would take place, because a part of the area in South Suriname will be flooded. Aquatic mammals, such as giant otters, may be particularly vulnerable.

Southeastern Suriname area is very important for large mammal species, because the area encompasses vast tracts of pristine forest and rivers. In fact, there are few places left on earth which are as pristine as Southeastern Suriname. Large mammal species have broad home ranges and can move freely across Southeastern Suriname in the absence of disturbance. The area is also connected to protected areas in Brazil to the south and to a national park in French Guiana to the east. This makes Southeastern Suriname part of a large wilderness area, which is important to sustain genetic diversity within large mammal species. Furthermore the area is important for the production of ecosystem services directly used by humans, such as water, food, medicines, recreation and lands of indigenous people. Since a large part of the rivers

in Suriname originate in this area, protecting Southeastern Suriname guarantees flows of fresh water which is used for economic activities such as transportation, hydro energy, agriculture and mining. The area also has a high potential for ecotourism, because of the beautiful pristine landscape and rich biodiversity, particularly of charismatic mammals.

It is important to keep following and collecting data on the plans of the national government and others regarding development of large-scale projects in southern Suriname. Roads, mining and hydro energy projects in southern Suriname will certainly increase the threats for large and medium-sized mammal populations in this area. More data on large mammal populations will be necessary to manage these possible future threats. Recommended studies include more camera trapping and a sustainability evaluation of wild bushmeat hunting to have better baseline data.

REFERENCES

Bennett, E. L., E. Blencowe, K. Brandon, D. Brown, R. W. Burn, G. Cowlishaw, G.

Davies, H. Dublin, J. E. Fa, E. J. Milner-Gulland, J. G. Robinson, J. M. Rowcliffe, F. M. Underwood, and D. S. Wilkie. 2007. Hunting for consensus: Reconciling bushmeat harvest, conservation, and development policy in west and central Africa. Conservation Biology 21: 884–887.

Bodmer, R. E, T. G. Fang, L. Moya, and R. Gill. 1994. Managing wildlife to conserve Amazonia forests: Population biology and economic considerations of game hunting. Biological Conservation 67: 29–35.

Cardillo, M., G. M. Mace, K. E. Jones, J. Bielby, O. R. P. Bininda-Emonds, W. Sechrest, C. D. L. Orme, and A. Purvis. 2005. Multiple causes of high extinction risk in large mammal species. Science 309: 1239–1241.

Cullen, L., R. E. Bodmer, and C. V. Pádua. 2000. Effects of hunting in habitat fragments of the Atlantic forests, Brazil. Biological Conservation 95: 49–56.

Diamond, J. 1989. The present, past and future of human-caused extinctions [and discussion]. Philosophical Transactions of the Royal Society of London. Series B, Biological Sciences (1934–1990) 325: 469–477.

Fa, J. E., D. Currie, and J. Meeuwig. 2003. Bushmeat and food security in the Congo Basin: linkages between wildlife and people's future. Environmental Conservation 30: 71–78.

Fa, J. E., C. A. Peres, and J. Meeuwig. 2002. Bushmeat exploitation in tropical forests: an intercontinental comparison. Conservation Biology 16: 232–237.

Hill, K., G. McMillan, and R. Fariña. 2003. Hunting-related changes in game encounter rates from 1994 to 2001 in the Mbaracayu Reserve, Paraguay. Conservation Biology 17: 1312–1323.

Lahm, S., A.and Tezi, J., P. 2003. Assessment of the Communities of Medium-sized and Large Arboreal and

Terrestrial mammals in the Rabi-Toucan Region of the Ngove-Ndogo Hunting Domain and Southwestern Loango National Park. Bulletin of the Biological Society of Washington, No. 12: 383–416.

Leader-Williams, N., S. D. Albon, and P. S. M. Berry. 1990. Illegal exploitation of black rhinoceros and elephant populations: Patterns of decline, law enforcement and patrol effort in Luangwa Valley, Zambia. Journal of Applied Ecology 27: 1055–1087.

Milner-Gulland, E. J., M. V. Kholodova, A. Bekenov, O. M. Bukreeva, I. A. Grachev, L. Amgalan, and A. A. Lushchekina. 2001. Dramatic declines in saiga antelope populations. Oryx 35: 340–345.

Milner-Gulland, E. J., and H. R. Akcakaya. 2001. Sustainability indices for exploited populations. Trends in Ecology and Evolution 16: 686–692.

Milner-Gulland, E. J., E. L. Bennett, and the SCB 2002 Annual Meeting Wild Meat Group. 2003. Wild meat: the bigger picture. Trends in Ecology and Evolution 18: 351–357.

Peres, C. 1990. Effects of hunting on western Amazonian primate communities. Biological Conservation 54: 47–59.

Purvis, A., P. M. Agapow, J. L. Gittleman, and G. M. Mace. 2000. Nonrandom extinction and the loss of evolutionary history. Science 288: 328–330.

Redford, K. H. 1992. The empty forest. Bioscience 42: 412–422.

Robinson, J., and E. Bennett. 2004. Having your wildlife and eating it too: an analysis of hunting sustainability across tropical ecosystems. Animal Conservation 7: 397–408.

Robinson, J. G., and E. L. Bennett. 2000. Hunting for Sustainability in Tropical Forests. New York: Columbia University Press.

Silva, J. L., and S. D. Strahl. 1991. Human impact on populations of chachalacas, guans, and curassows (Galliformes: Cracidae) in Venezuela. pp. 37–52 In: Robinson, J. G., and K. H. Redford (eds). Neotropical Wildlife Use and Conservation. Chicago: University of Chicago Press.

Silvius, K. M., R. E. Bodmer, and J. M. V. Fragoso. 2005. People in Nature: Wildlife Conservation in South and Central America. New York: Columbia University Press.

Turvey, S., and C. Risley. 2006. Modelling the extinction of Steller's sea cow. Biology Letters 2: 94–97.

Chapter 13

Primates of the Grensgebergte and Kasikasima region of Southeastern Suriname

Arioene Vreedzaam

SUMMARY

Six of the eight primate species known from Suriname were recorded during the RAP survey. These included the black spider monkey (*Ateles paniscus*), red howling monkey (*Alouatta maconnelli*), bearded saki (*Chiropotes sagulatus*), brown capuchin (*Cebus paella*), squirrel monkey (*Saimiri scuireus*), and golden handed tamarin (*Saguinus midas*). The large bodied species (black spider monkey, red howling monkey) were present in relative abundance indicating sustainable hunting practices by local communities. Lack of records of the white faced saki and wedge capped capuchin do not necessarily mean that they are not present. These species are quite difficult to spot due to rarity and elusiveness. From a conservation perspective both sites can be considered to have healthy populations of monkey species.

INTRODUCTION

The greatest floristic diversity is found in tropical regions, particularly in the forests of Western Amazonia and Borneo (Wright 2002). Within these areas plants and animals engage in varying degrees of interaction which impact seed dispersal, seedling recruitment, and ultimately the maintenance of rainforest tree diversity. In terms of fruiting tree diversity, birds, bats, rodents, and primates play a key role in maintaining this diversity.

In the forests of Suriname, frugivorous primates are regarded as primary seed dispersers for many fruit bearing tree species: black spider monkeys, red howling monkeys, and capuchins disperse a wide array of seeds from various species. Seed predators (white-faced saki, bearded saki) on the other hand, consume a wide variety of seeds from different plants which reduces the predator load on any one plant (Norconk et al. 1998). Additionally, there is evidence that seed predators enhance seed germination by removing pericarp or seed coating (Norconk et al. 1998).

Assessment of the primate diversity in the Grensgebergte and Kasikasima regions is essential, especially considering the fact that large bodied primate species (black spider monkey, red howling monkey, brown capuchin, bearded saki) form an important part of the diet of the local communities, the Trio and Wayana people. The presence of these primate species is not only an indicator of forest health, but also indicates sustainable extraction practices by local communities.

METHODS

At both sites, the Juuru Camp and Kasikasima, transects were utilized to conduct primate surveys. When the weather permitted, surveys were conducted in the morning and afternoon whereby audio and/or visual cues were used to record the presence of primate species. The following data were collected during each survey: species, audio and/or visual detection of species, observer to animal distance, animal to trail distance, time of recording, and GPS point. Additional behavioral data were also collected regarding animal reaction to the observer.

At the Juuru Camp I used three transects to conduct surveys: from base camp to the Northeast up and over the nearest hill, from the top of the hill down the slope to the North, and the Brazil trail which heads to the South. The total distance of all trails is 5.14 kilometers. Aside from the Brazil trail, all trails ran over very uneven hilly terrain and through varying forest types. Along parts of these trails elevation change was quite drastic, going from 277 m to 538 meters.

At the Kasikasima site three trails were also utilized for surveys: from base camp to the METS tourist lodge (North), from base camp towards the METS tourist trail (West), and from the METS tourist trail to and up the Kasikasima Mountain (Southwest). The total distance of all trails is 6.9 kilometers.

RESULTS

Table 13.1 presents a list of the primate species recorded during the RAP survey. Six of the eight primate species known from Suriname were recorded during the survey.

Juuru Camp site

Five of the eight Suriname monkey species were recorded: the black spider monkey, red howling monkey, brown capuchin, squirrel monkey, and the golden handed tamarin. In terms of visual encounter frequency, the golden handed tamarin and the squirrel monkey had a higher encounter frequency than the other three monkey species. Black spider monkeys and red howling monkeys were more often heard than seen, while the brown capuchin monkeys were encountered briefly on two occasions.

Kasikasima site

Species recorded at the river camp site were the black spider monkey, red howling monkey, bearded saki, brown capuchin, and the golden handed tamarin. Along the foot of Kasikasima Mountain black spider monkeys, red howling monkeys, and golden handed tamarins were spotted.

The white faced saki and the wedge capped capuchin were not spotted at either site.

Table 13.1. The eight primate species known from Suriname, with records of those documented during the RAP survey.

Common Name	Species	Juuru Camp	Kasikasima site
black spider monkey	*Ateles paniscus*	x	x
red howling monkey	*Alouatta maconnelli*	x	x
bearded saki	*Chiropotes sagulatus*		x
white faced saki	*Pithecia pithecia*		
brown capuchin	*Cebus apella*	x	x
wedge capped capuchin	*Cebus olivaceus*		
squirrel monkey	*Saimiri scuireus*	x	
golden handed tamarin	*Saguinus midas*	x	x

CONSERVATION RECOMMENDATIONS

The following statements can be made regarding primate conservation in the Grensgebergte and Kasikasima Mountain areas:

- Preliminary results indicate that the large bodied species (black spider monkey, red howling monkey) are present in relative abundance. They were either spotted or heard on a regular basis at both sites. Since these two species are the most hunted by local communities (pers. comm.), this indicates sustainable hunting practices by these communities.

- The absence of the bearded saki, white faced saki, and wedge capped capuchin at the Grensgebergte site does not necessarily mean that they are not present. These three species are quite difficult to spot due to rarity and elusiveness. The bearded saki however, was spotted once at the Kasikasima site. More data need to be collected regarding these "missing" species.

- Absence of the squirrel monkey (a common monkey species) at the Kasikasima site is a result of the lack of suitable habitat for this species. They prefer liana forest and low riverine forest types, both of which were not present at the Kasikasima site.

- From a conservation perspective both sites can be considered to have healthy populations of monkey species.

REFERENCES

Norconk M.A., Grafton B.W., and Conklin-Brittain N.L. 1998. Seed Dispersal by Neotropical Seed Predators. American Journal of Primatology 45:103–126.

Wright S.J. 2002. Plant diversity in tropical forests: a review of mechanism of species coexistence. Oecologia 130:1–14.